Principle Concepts of Technology and Innovation Management:

Critical Research Models

Robert S. Friedman
New Jersey Institute of Technology, USA

Desiree M. Roberts
State University of New York-Empire State College, USA

Jonathan D. Linton
University of Ottawa, Canada

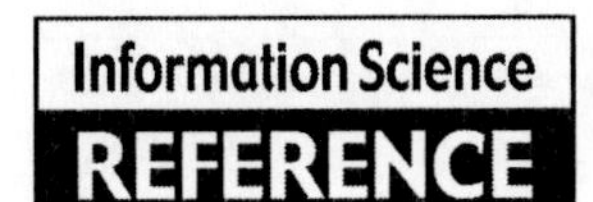

INFORMATION SCIENCE REFERENCE

Hershey · New York

Acquisition Editor:	Kristin Klinger
Managing Development Editor:	Kristin Roth
Senior Managing Editor:	Jennifer Neidig
Managing Editor:	Jamie Snavely
Assistant Managing Editor:	Carole Coulson
Copy Editor:	Maria Boyer
Typesetter:	Chris Hrobak
Editorial Assistant:	Heather A. Probst
Cover Design:	Lisa Tosheff
Printed at:	Yurchak Printing Inc.

Published in the United States of America by
IGI Publishing (an imprint of IGI Global)
701 E. Chocolate Avenue
Hershey PA 17033
Tel: 717-533-8845
Fax: 717-533-8661
E-mail: cust@igi-global.com
Web site: http://www.igi-global.com

and in the United Kingdom by
IGI Publishing (an imprint of IGI Global)
3 Henrietta Street
Covent Garden
London WC2E 8LU
Tel: 44 20 7240 0856
Fax: 44 20 7379 0609
Web site: http:/www.eurospanbookstore.com

Library of Congress Cataloging-in-Publication Data

Principle concepts of technology and innovation management : critical research models / Robert S. Friedman, Desiree M. Roberts, and Jonathan D. Linton, editors.
p. cm.
Summary: "This book is a reference guide to the theory and research supporting the field of Technology and Innovation Management"--Provided by publisher.
Includes bibliographical references and index.
ISBN 978-1-60566-038-7 (hardcover) -- ISBN 978-1-60566-039-4 (ebook)
1. Technological innovations--Management. 2. Organizational change. 3. New products. I. Friedman, Robert S., Ph. D. II. Roberts, Desiree M. III. Linton, Jonathan D.
HD45.P735 2009
658.5'14--dc22

2008010317

British Cataloguing in Publication Data
A Cataloguing in Publication record for this book is available from the British Library.

Table of Contents

Foreword

This book is essential reading for any reflective practitioner, researcher in the field, or master's student basing their professional career on any of the nine topic areas in Technology and Innovation Management that this book covers. This is the first work since Dr. Adler's effort in the early 1990s to discuss the many subfields of Technology and Innovation Management (TIM) as a multi-discipline, multi-dimensional academic subject, and it is about time! The authors provide a concise discussion of the seminal work in the top subfields of this rapidly growing area.

The authors have split the effort into three main sections and two appendices that, when coupled with thoughtful introduction and summary sections, provide exceptional value. The two appendices with the most cited papers and words are exceptionally useful for the research and practitioner communities. The executive summaries, introductory comments, and epilogue are a must read.

The authors provide the readers with the fundamental ideas that have come to form the base of information that academicians, researchers, and practitioners involved with TIM should know. The authors have categorized nine topic areas into three main sections. First the authors artfully weave "R&D Process Models," "Technology Development and Innovative Practice," and "Social Influence and Human Interaction with Technology" into a section they title **Processes, Strategies, and Development.** Next they integrate both seminal and interesting new thought in four major areas of "Diffusion and Innovation: An Organizational Perspective," "Knowledge and Change in Organizations," "Organizational Innovation Strategy," and "New Product Development" into a section they title **Innovation, Influence, and Diffusion.** Finally they cover **Technology and Management Information Systems** through their discussion of two sub-areas of the field: "Information and Communication Technology Management" and "Open Source and Software Development Innovation."

The authors have done an exceptional job of integrating the new books on innovation, technology, and management that appear frequently, and the older chestnuts such as Everett M. Rogers' *Diffusion of Innovations* which are reissued on a regular basis. What Friedman, Roberts, and Linton have done is to crystallize nine essential TIM concepts from a historical perspective, then synthesize the main ideas of relevant topics for academic researchers, students of related disciplines, or practitioners interested in what theorists have developed over the past half-century. A truly exceptional and useful effort.

Steve Walsh, PhD
Albuquerque, NM
January 2008

Preface

This book is a reference guide to the theory and research supporting the field of technology and innovation management (TIM). Through the presentation of the seminal ideas articulated in the discipline's fundamental texts and most widely cited journal articles, we present pertinent information—from basic definitions to some of the most advanced theoretical knowledge and empirical data of the discipline's pioneers and practitioners—of import to academics, innovators, and managers alike. Citation analysis was performed to identify the articles most referenced, and based on those results the key ideas of the articles were organized into thematic and sub-disciplinary groupings that yielded the sections and chapters of this book. The selection of frequently cited papers at least partially reflects the relative size of the different research communities, and there is a tendency for works to be cited by researchers within the associated traditional discipline as opposed to researchers who also study technology and innovation management from the perspective of an alternative traditional discipline.

In Chapter I, we give primary consideration to books that are critical to the understanding and study of technology and innovation management, then provide a discussion of what technology is, how innovation is now defined—anything that involves a change that is new or novel to the individual or organization involved—considering *new* as the pivotal word. Chapter I contains a discussion of the interaction of innovation and technology with management. Given the academic atomization of different traditional fields that study technology and innovation management, we consider the most frequently cited papers in all fields contributing to TIM with the intention of closing some of the gulf that separates these communities.

Chapter II, with its focus on research and development (R&D) processes and models, begins with a discussion of the level and effectiveness of R&D managers regarding budgeting, an econometric model of the relationships of R&D with financing decision making, the different stresses and influences that investors and consumers place upon the R&D process, and the effects of management structure and diversification on investment by external capital markets in R&D firms. Diversification strategies that affect R&D and their impact on the selection of internal or external R&D sources when technological changes affect the locus of R&D expertise is also discussed, as are the roles of knowledge transfer, both within R&D organizations and among various technology stakeholders. We conclude the chapter with a look at "absorptive capacity," particularly some of the key elements of knowledge transfer.

There are three dominant themes related to organizational innovation strategy that run through Chapter III: the rate and nature of change; attitudes, behaviors, and strategic change; and the role of research in organizational strategy. We begin with the interaction between technology and organizational structure to uncover how this kind of interaction affects how organizations function. A study that contrasts the size of firms to their attitudes toward innovation is presented, followed by a discussion regarding the continuum of incremental-to-radical innovation by taking a very close look at the innovative process—from manufacture to end-user sale. The role of research in strategic change is also a subject of this chapter, with a goal of helping determine the role of 'competence' in an industry's research. Component and architectural competence are examined and explain the nature of variance in research productivity, as is the nature of competition in organizations undergoing strategic change.

Chapter IV concerns new product development, with articles that address the internal processes that assist or hinder development and factors that contribute to the success or failure of a new product, including its performance and diffusion. We begin with the steps that affect the development process and determine how modifying a step-wise structure improves process performance, and continue with a discussion of the tensions and trade-offs that occur among different functional areas and how they affect innovative product development. We then review contemporary new product growth models as a basis for understanding recent diffusion models of new product acceptance. A sociological analysis of people's communicative behavior suggests that the objective of a diffusion model is to illustrate the increases in the scope of adopters and predict the nature of the development of an ongoing diffusion process. This chapter also contains a look at empirical studies of product development that focus on the development project as the element of analysis in order to provide a model of factors that contribute to the success of new product development.

Chapter V, on technology development and innovative practice, begins with seminal work on routinization and how social structures of organizations affect technological development and innovation, approached from both statistical and sociological perspectives. We also look at the changing definition of "innovation" through factors of variability and quantitative methods. The discussion continues with the inclusion of theoretical constructs for innovation that identify variables of structural differentiation and complexity that affect this domain. Then, researchers describe how a combination of technology sources, user requirements, and potential technology appropriation affect how we understand technical change and the structural relationships between technology and industry. The term "transilience" is highlighted to indicate a set of categories of technological change that is aligned with evolutionary developments, altered by varying managerial environments.

In Chapter VI we discuss how information that supports innovation flows throughout an organization, the construction and effects of team composition, the innovative process that teams employ, and the development, implementation, and evaluation of systems used to manage the flow and distribution of information. Research indicates that effective communicators rise as a result of their willingness to engage information. We also discuss why innovative processes require the development of effective information networks, confirming how important it is for successful innovation that there exist effective external and internal communication networks, and that individuals collaborate to share information. Team composition is another theme of this chapter, with researchers suggesting that certain demographic factors affect a team's ability to be innovative, but resource diversity—including communication ability—is ultimately essential to innovation, as are corporate executives' abilities to understand and adapt to the fact that the innovation environment is filled with surprise.

Chapter VII introduces the seminal literature addressing technological diffusion, innovative product diversification, and the organizational strategies and constraints that firms face when introducing and adopting new technologies and innovative management strategies, drawing critical distinctions between the processes undertaken by rational adopters of inefficient technologies and the conditions that promote the irrational rejection of efficient innovations. Chapter VII also addresses diversification and organizational structure by locating a theoretical basis for the identification and validation of factors that influence diversification innovation adoption strategies. The important concept of structuration provides an alternative conceptualization of the role of technology, focusing on the theory's social and historical substrata to provide an explanation of how we might rethink the roles of technology in organizations.

The focus of Chapter VIII is on the role of knowledge in the operation of organizations, and it consists of two main thrusts: the effects of knowledge (accrual, dissemination, and implementation) on organizational change, and more specifically, the manner and effects of knowledge transfer within and among firms conducting innovative product design and development. We look at the importance and processes of knowledge coordination within a firm's administrative hierarchy, the role of radical change on the theory of neo-institutionalism, and how one identifies and exposes organizational capabilities in the face of organizational structures that promote management practices having the potential to stifle innovation rather than institute and nurture change. Researchers investigating knowledge transfer offer reasons for and processes by which competing firms exchange organizational knowledge, finding a range of distinguishing characteristics between the subject matter and substance of inter-organizational arrangements and the organizational structures and complexities of those firms. The chapter also discusses the symbiotic relationship between technological innovation and its adaptation into the organizational environment, how research is organized in science and technology sectors to point out how interrelated and complex their activities are, and how knowledge can be viewed as an instrument of organizational change.

Chapter IX, on information and communication technology management, first presents the development of research concerned with behavior—specifically, attitudes and decision behavior in the early realm of management information systems—then shifts focus toward methodologies and practices of MIS development and their implementation.

The third and concluding section of this chapter follows the progression of information and communication technologies from the mid-1980s to present as it shifted focus to the individual customer—as development partner, and as arbiter of product design and modeling. The well-known theory of reasoned action (TRA) and technology acceptance model (TAM) are discussed, as is an approach to information systems development from a strategic and organizational (as opposed to a user-based) vantage point: adaptive structuration theory (AST). Other research introduces situated practice as a methodology to understand the relationships between organizational change and IT, and an alternative to established perspectives such as planned change, technological imperative, and punctuated equilibrium. We conclude this chapter with further discussion of the Technology Acceptance Model, but with consideration given to additional variables and their effects on perceived usefulness and perceived ease of use of information technologies.

Chapter X concerns open source and software development innovation, with research that addressed the development of software and the challenges it poses to commercial concerns, as well as specific situations in which management and innovation theory is responsive to non-proprietary software development. Researchers introduce advances in theory to aid software project management and discuss risk management as it pertains to software development projects. Knowledge management in software process innovation management environments is addressed, and the chapter concludes with an overview of strategies for organizations seeking to meld proprietary and open source methodologies and management styles with established theories of appropriability and adoption as they pertain to software development.

We conclude the book with a look to the future, but one in concert with the underlying theme that holds together the research considered in previous chapters: the tension between the old (current routine) and the new (innovation). While it is not possible to state with any certainty which recent research will be considered seminal work several decades from now, it is possible to give insights into current trends in research and to project these out into the future. We employ a methodology similar to the one identified in the seminal work discussed in the earlier chapters. Identifying the most cited articles in TIM through the Science Citation Index (SCI) and Social Science Citation Index (SSCI), a count of individual and pairs of words was conducted to identify the frequency of occurrence, then the list was compared to a similar list of words from all management journals listed on both indices. We identified the following areas as trendsetters for future research in TIM: new product development, technology transfer, supply chains, network or relationship-related concepts, and new emphases and approaches to technology transfer research. There is increased attention to technology and innovation management in emerging and developing economies, and it is likely that there will be a reemergence in discussion of appropriate technology (as opposed to high tech).

Acknowledgment

The authors would like to acknowledge the contributions of W. George Biggar, senior IT specialist at Rensselaer Polytechnic Institute, whose efforts made the underlying reading notes possible by taking responsibility for converting the 9,000 pages into a format compatible with Dr. Roberts' adaptive software.

We would also like to dedicate this book to people whose support, in different ways, helped in its development. From Jonathan Linton to his family, "The Four Musketeers": Jacob, Ariel, Jordan, and wife Helena; from Desiree M. Roberts to her husband, Thomas H. Roberts; and from Robert S. Friedman to his mother, Taube Friedman, in memoriam.

Section I
Processes, Strategies, and Development: Executive Summary

The chapters in this section are related by their foci on models and strategies for innovative product development and management processes. In Chapter I, primary consideration is given to the fundamental books, authors, and theories that support a critical understanding of technology and innovation management. Principle ideas and base-line definitions of technology and innovation precede a discussion of the interaction of innovation and technology with management. Chapter II addresses R&D processes and models through the lens of economics and finance, financial decisions hypothesis and development models that help managers understand the relationships of R&D with financial decision making, the stresses and influences that investors and consumers place upon the R&D process, and the nature of management structure and diversification as they are affected by external capital markets in R&D firms. Diversification strategies that affect R&D are also discussed in terms of the positive relationship between the level of R&D intensity and the level of business dominance. Chapter II also addresses the role of transaction costs and their effect on the selection of internal or external R&D sources when technological changes affect the locus of R&D expertise; the roles of knowledge transfer, both within R&D organizations and among various technology stakeholders; methods of assessing the level of contribution that different research groups of a specific content area contribute to their field's knowledge base; and how the information relationship between research and development and marketing integration is fundamental to a firm's business strategy. Absorptive capacity, or how individuals and firms assimilate and use information to fuel innovation, is a key concept of this section, as is the nature and degree to which technological innovations are based on academic research and the time innovators' expend in engaging academic research and industry's subsequent use of their results.

With Chapter III, the literature centers on innovative practice supporting technological development, particularly on routinization and how the social structures of organizations affect technological development and innovation. Readers will discover how path analysis is used to study the nature and effects of organizational variables on innovative practice, how quantitative methods are shown to be increasingly powerful tools for identifying the nature of innovation and technology development, and how variables of structural differentiation and complexity affect this domain. Sectoral pattern analysis is used to describe how a combination of technology sources, user requirements, and potential technology appropriation affect how we understand technical change and the structural relationships between technology and industry. Another key term introduced in this section is "transilience," or a set of categories of technological changes aligned with evolutionary developments and altered by varying managerial environments. Cyclical models of technological change, evolutionary-based views of technological change shaped by customer demands, and sociologically oriented sets of ideas demonstrate the strength of identifying patterns of continuous changes and moments of discontinuity in technologically innovative environments.

The section concludes with a discussion of how the information that supports innovation flows throughout an organization, the effects of team composition on innovative practices, and the nature of innovative processes that manage the flow and distribution of information. The concept of special boundary roles is addressed, as are the conditions necessary to move an organization from a single-minded focus on productivity to one that facilitates innovation. Readers will find discussion on the benefits of information sharing, how information regarding innovative processes entails the development of effective information networks, and how team composition affects its ability to be innovative. Also addressed are the function and significance of the values held by an organization's elite group in terms of the innovative processes in an organization, as well as how information technologies affect management strategies and how these strategies are disseminated throughout a firm.

Chapter I
Introduction to the Field of Technology Innovation Management

INTRODUCTION

This book differs from other academic works on the management of technology and innovation because it focuses on the seminal research of the field. Such work continues to be returned to by many authors over time because it supplies information considered to be core and foundational in nature. Consequently, the focus of this book is on older work that appears to be of increasing relevance over time and newer work that has quickly become highly influential. For the specialist practitioner interested in a specific technology or the academic who is interested in innovation from the perspective of a specific traditional discipline, this book will provide you with a strong foundation that cuts across traditional fields and boundaries. With the foundational knowledge in place, readers have a solid base over which to place the specialist knowledge that is of importance to them. Although the focus of the book is on foundations, the section on technology and management information systems offers additional insight into MIS, which many information systems professionals, universities, and professional organizations consider to be an independent discipline of increasing importance to fields that use information to develop and alter business policies and procedures. The final chapter focuses on the future of technology innovation management. By conducting a textual analysis of recent research from the top specialty journals in technology innovation management, we offer the reader sufficient information to consider what topics and directions recent research in the technology innovation management specialty is taking.

The first step in considering the field of technology and innovation management (TIM) is to offer some very basic definitions to ensure that it is clear what is meant by the authors when certain terms such as 'technology' and 'innovation' are used. Over the last few years, these terms have been overused and in some cases abused. Because these terms are fashionable and have a positive connotation, they tend to be used somewhat less than sparingly. Consideration of technology often focuses on how science is different from engineered or technology-based products. Stokes (1997) considers the difference of understanding for the purpose of increasing knowledge (science) vs. understanding for the purpose of application or problem solving (technology). He suggests that it is possible to make a contribution to knowledge that offers no practical application. For example, Bohr's model of the atom is important to our understanding of science, but it does not contribute to the development of products. Such a discovery is termed as basic science.

At the other extreme, some advances result in the development of product, but no increase in knowledge occurs; this is applied science or technology. An example is the development of the light bulb by Edison through experimentation with a huge number of materials, until he found one—tungsten—that performed satisfactorily.[1] These two examples offer the extreme points of the spectrum. Technology does not need to increase knowledge, but must offer some applied benefits.[2] Science involves an increase in knowledge or understanding, but does not need to offer applied benefits. In many cases, however, scientific discovery offers both advances in knowledge and one or more applied benefits. For example, the process of pasteurization offers both applied benefit and scientific understanding regarding the presence and existence of micro-organisms.[3] Having given an initial consideration of what technology is, innovation is now defined. Innovation is considered here as anything that involves a change that is new or novel to the individual or organization involved.[4] The critical idea here is *new.* Anything that is not new to the organization falls under the heading *change management.*[5] Having considered the meaning of innovation and technology, we now consider the interaction of innovation and technology with management.

Technology and innovation management is challenging to consider from the perspective of a field, since it is outside of and crosses the boundaries present in the traditional disciplinary structure used by most academics and universities. Consequently, there are communities within different traditional fields that study technology and innovation management. However, these communities study technology innovation management from very different perspectives. Many of these differences in perspectives are fundamental and foundational. As a result, these communities are often isolated from each other, with researchers' findings having much more in common with the associated traditional discipline than with researchers studying technology and innovation management from a different perspective.

By considering the most frequently cited papers in the field of technology innovation management, the gulf that separates these communities is partially spanned. However, the selection of frequently cited papers at least partially reflects the relative size of the different communities and the tendency of the work to be cited by researchers within the associated traditional discipline as opposed to researchers who also study technology and innovation management, but from the perspective of an alternative traditional discipline. The consideration of books critical to the understanding and study of technology and innovation management is an ideal place to address at least part of this concern. Much of the journal literature considered in this book is written from the perspective of the firm, often the large firm. This is just one example of a community that considers elements of technology and innovation management. The intense consideration of the firm, in the journal literature, is not only a reflection of the importance of this unit of analysis, but the perspective of schools of business administration and management. If we move back in time, as we will for consideration of critical texts, only a relatively short time must be traversed for us to minimize and then completely eliminate the firm perspective. Large firms, as a common mode of organization, came about in response to opportunities created around the time of the industrial revolution. Prior to this time, large firms such as the Hudson Bay Company (founded in 1670) or the Dutch East India Company (founded in 1602) were very rare.

The first discussion of technology is typically credited to *The Republic,* 360 B.C. (Plato, 1987). Plato discusses the term *techne.* The usage of the term suggests that *techne* is the *art of work.* In other words, *techne* is the enveloping term representing the skills that an artisan develops or has developed (skills that other people lack) that enable him or her to excel in the non-codified creativity or art that is related to their field of endeavor (such as the making of a vase or piece of jewelry). The concept of *technology* relating to the art of work, as opposed to the science of work, is quite alien to current usage of the word *technology.* Today the word *technology* has a tight linkage to the idea of science. However, if one considers that the codification of knowledge was limited millennia ago and pursuit of knowledge lacked the structure of today, then *techne* as art is an unsurprising definition.[6]

Having identified the start of our journey through seminal thought in technology and innovation management, it is now important to move both forward through time and across the disciplines that both inform and separate the field. Since time and disciplinary perspective are both critical to understanding the richness of technology innovation management, from this point onwards these two considerations will be taken.

In the earlier years, the field of economics tends to dominate the writings in technology innovation management. The economists in some cases use a regional, national, or transnational economy as a unit or level of analysis. However, in other cases economists have used technology as a unit or level of analysis. On occasions that technology has been considered as a focus, the economists have considered technology not as an end, but as a means of better understanding changes in economic growth and/or economic structure (Kondratieff, 1984; Schumpeter, 1934, 1939a, 1939b, 1952; Kuznets, 1979).

THE TECHNOLOGY PERSPECTIVE IN INNOVATION

To find central consideration given to a technology and its development, one must turn to the field of history and the sub-field of the history of science. Much of what has been written considers the development of technology from the perspective of the work of an individual or an organization. For example, books such as *The Manhattan Project* (Beyer, 1991) or *Copies in Seconds* (Owen, 2004) tend to focus on personalities and the drama around these personalities. Such books often end with the death of a main character or the completion of some critical event, while a book that focuses specifically on the history of a technology—such as *Salt* (Kurlansky, 2002) or *Medieval Technology and Social Change* (White, 1965)—concerns the evolution of a technology including critical experiments, discoveries, and conceptual breakthroughs leading to improvements in the scope, effectiveness, and efficiency of a technology. This sort of assessment will end with the technology either reaching maturity or being dominated and replaced by an alternative technology system.

Books that focus on the human events surrounding science and technology make for interesting reading and place people in control. Books with a technology-centric focus also offer interesting reading, but place the technology as the dominant force shaping people's lives and environments. In these books, people often have little control of events. The most extreme examples of technology-in-control are the nightmare scenarios presented in movies such as *The Terminator* (Cameron, 1984) or *The Matrix* (Wachowski & Wachowski, 1999), or social commentaries such as *Modern Times* (Chaplin, 1936) or *Silent Spring* (Carson, 1962). In addition to fictional accounts, the unanticipated effects of technology are described in studies such as Hightower[7] (1972) or Parfit[8] (2000). The nightmarish aspects of the possible unanticipated effects of technology are addressed by ethicists, scientists, and policymakers[9] in an attempt to avoid such problems. Consideration of the possible nightmare scenarios or the more general impacts of technology and innovation are difficult to predict. Attempts at system modeling and the use of expert panels have led to some notable failures.[10] It is fair to state that the field of predicting future impacts of technology and innovation on a region, country, or internationally is a field that is still wide open for development.

The most successful attempts at forecasting the effects of technology result from the work of Kondratieff (1984). Unfortunately, Kondratieff was at the wrong place at the wrong time—Soviet Russia. His work was considered to be politically unsound, and he spent many years in a gulag as a result of his alternative thinking.[11] Kondratieff proposed the concept of long waves. He theorized and demonstrated empirically that the economies of several different countries seem to go through waves of 60 years in duration (see Figure 1). These waves involved times of economic prosperity and expansion, as well as times of economic dislocation and disruption. Curiously, the timing of military conflicts appears to fit well into his theory of long waves or cycles. Kondratieff indicates the relation between major technological advances and the cycle, but does not state the implied cause-and-effect relationship so forcefully. Kondratieff lists technologies that he believes to be responsible for cycles, but offers no guidance on identifying a technology that can be the basis of a Kondratieff wave or how much time elapses between scientific discovery, the introduction of a technology, and the onset of a Kondratieff wave.[12] It is not clear if all of the abovementioned omissions were intended, either due to local political necessity or because Kondratieff was not fully satisfied with the proof of the relation.[13] Another important question that emerges from Kondraitieff's work is the relationship between conflict and technology. In other words, does the presence of new technology increase the likelihood of war, or do wars accelerate technological discovery, advancement, and diffusion? Or is there an interrelationship between war and technology—does one lead to the other?

The concept of the long wave is further considered and developed. Schumpeter's theoretical contribution is based on his recognition of the disruptions and benefits associated with entrepreneurship (Schumpeter, 1934, 1952). The entrepreneurs are

Figure 1. Depiction of a Kondratieff Wave (based on The Economist, 1999)[14]

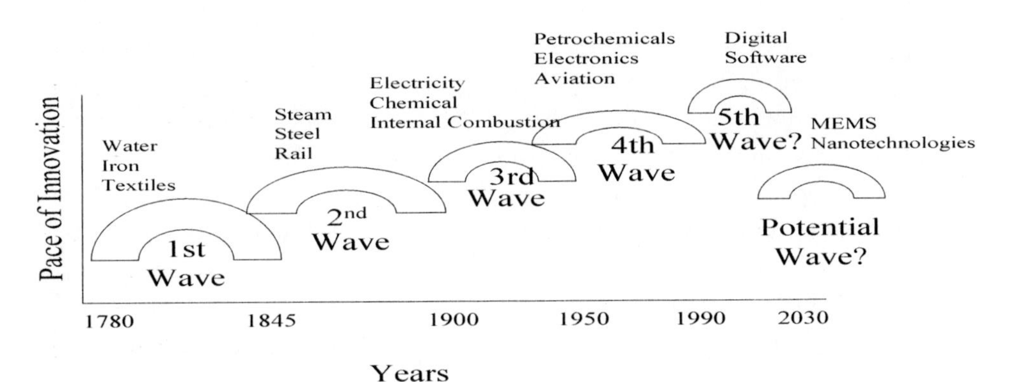

individuals who *carry out*[15] innovations. In his less known work on *Business Cycles,* Schumpeter (1939a, 1939b) discusses economic cycles being a function of short-, medium-, and long-term cycles. The short waves—Kitchins—consist of intervals of about a 40-month duration identified by Crum (1923), based on data on interest rates for commercial paper, and by Kitchin (1923) based on wholesale prices and bank data. A medium wave—Juglar's[16]—has a duration of about 10 years. Finally, there is the long wave—Kondratieff—with a duration of about 60 years. From the perspective of technology innovation management, consideration of only the long wave is critical. However, it is important to recognize that Schumpeter and many others contributed to the rich literature on economic cycles in the early 20th century.[17]

Schumpeter identifies what he believes to be the approximate starting point and technological basis of three long waves—the industrial revolution 1780-1842, steam and steel technology 1842-1897, and a wave based on electricity, chemistry, and motors starting in 1898. Such a progression suggests that this final Kondratieff wave ends sometime between the end of the Second World War and 1950. Others have proposed additional Kondratieff waves, for example: a proposed fourth wave of oil, electronics, aviation, and mass production, and a fifth wave of semiconductors, fiber optics, genetics, and software (The Economist, 1999). Schumpeter suggests that during each Kondratieff wave, idea generation and implementation activities deal with the application of different opportunities enabled by the innovations driving the current Kondraiteff wave. By the end of the wave, these opportunities cease to exist and the entrepreneurs focus on opportunities enabled by the next Kondratieff wave's innovations.

The suggestion that entrepreneurs focus their activities on the opportunities that are enabled in each wave informs much of the thought of Nobel laureate Simon Kuznets that relates to innovation (1979). Kuznets estimates that 75% of U.S. economic activity is a result of innovation.[18] He adds to this theory by suggesting that changes in social structure are often required to allow the technological innovation to generate economic growth potential. Need for changes to social structure, along with recognition of individual opportunities for technological innovation and development, provide a basis for understanding why the characteristics that govern their potential result in Kondratieff's long waves take so long to be completed. In identifying the importance of social change, Kuznets identifies a difference between technological and social innovations, and suggests that technological innovations result in changes to social structures (social innovations) and that technological innovations are dependent on social innovations for enabling or providing access to a significant part of their economic benefit. In identifying innovation as a major factor in economic activity, and through his support of long wave theory based on a single or small number of critical innovations, he suggests that innovations should be divided up into those that are fundamental and those that offer additional or small improvements onto existing innovation.[19] Examples of social innovations include: division of labor into specialized functions, formation of banks, and founding of stock markets.

Prior to the industrial revolution, production was based on specialized artisans and trade guilds. Division of labor (Smith, 1776) is a social innovation that supports the move from artisan-based production to industrial production. While the specialization of labor into small units supports industrial production, in the absence of corporations it would be difficult, at best, to have the appearance of large production facilities and private railroads. The corporation is clearly a social innovation that was made necessary by both the industrial and iron ages, and has allowed the extraction of value out of these fundamental technologies. In the presence of the corporation, but also in the absence of large amounts of easily accessible funds, many of the benefits of innovation would either be too expensive to obtain, or be obtained only over a much longer period of time as a firm accumulated sufficient growth. Consequently, social innovations related to banking and the stock market were required to allow for the full development of innovations that were significant in initial cost and scope—such as privately financed railroads. In the absence of such innovations, it is likely that much of the economic benefit would not have been obtained.

Having introduced and demonstrated the relation between social and technological innovation, we consider Kuznets' last major concern relating to fundamental innovation—diffusion. Kuznets characterizes diffusion of a fundamental innovation as measuring how much of the value of the fundamental innovation has been extracted. In other words, a fundamental innovation would be fully diffused once the increase in economic value associated with the innovation has resulted in an increase in other factors of production, such as more people using the innovation due to population growth. So for Kuznets, diffusion measures the value produced by the fundamental innovation as: (1) more individuals or firms utilizing the fundamental innovation, (2) more applications of the fundamental innovation being determined, and (3) improvements occurring with the fundamental innovation. Kuznets recognizes that the critical problem with this definition is the determination of what constitutes a fundamental innovation. If steel and the transistor are fundamental innovations, does this relegate the personal computer to an incremental improvement of these fundamental innovations?

While there might be practical challenges to implementing and measuring this definition of diffusion, consideration of this definition is critical in areas such as technology strategy and policy. If a country or a firm focuses on one or more fundamental technologies as a basis for economic growth and profit, it is more likely to be a long-term success.[20] This realization is reflected in the focus of many developed and emerging economies currently focusing on research that may be the basis of fundamental innovations (e.g., Hung & Chu, 2006; Winickoff, 2006; Lane & Kalil, 2005; Kishi & Bando, 2004). One last point that is worth making here is that Kuznet's use of the term diffusion is quite different from the common current use, which reflects that percentage of the potential market that is using a product (see Rogers, 1995). Having considered technological innovation as a unit of analysis, we turn to the more current focus on individual, group, firm, industry, and market. Throughout the following chapters, we address these perspectives through the consideration of the most frequently cited academic articles. Consequently, consideration is limited in this chapter to important contributions that are based in books, rather than journals.

BRIDGING THE TECHNOLOGY AND PRODUCT PERSPECTIVES

Having considered innovation from a strictly technological perspective, we move to the interaction of the technology perspective with the different but related perspective of the product. Technological learning and innovation express their value through the creation of new products or the improvement in one or more dimensions of existing products. Technology has been described with terms such as 'platform' (Sanderson & Uzumeri, 1996) to indicate its ability to create or improve a variety of products, and the term 'trajectory' (Foster, 1986) to indicate the dynamic nature of these changes. It is critical to recognize that phenomena progress from a state where little is understood to one of great knowledge, and as this progression occurs, our ability to extract the benefits that are desired from the technology increases. This relationship has been described by a number of different authors. Bohn (1994) offers a model to explain different levels of learning.[21] The importance of focus on a technology to obtain improvements is considered directly by Foster (1986) and indirectly through the learning curve literature.[22] Foster focuses on the positive impact that R&D has to improvements in the performance of a technology (see Figure 2). He suggests that at first, a tremendous amount of effort is required to obtain a small improvement in performance—this corresponds to the first stage of Bohn's (1994) model: 'know-nothing'. At some point, however, a given amount of R&D on a technology will lead to increasingly larger improvements. Finally, as we approach full knowledge of a technology, we reach the physical limits of the technology. As we approach the technology's physical limits, the improvements obtained for a given level of R&D drop off quickly. A critical, but sometimes overlooked, contribution of this work is the recognition that R&D typically focuses on increase in performance on a finite, often single number of desired benefits.

Figure 2. Illustration of a technology trajectory

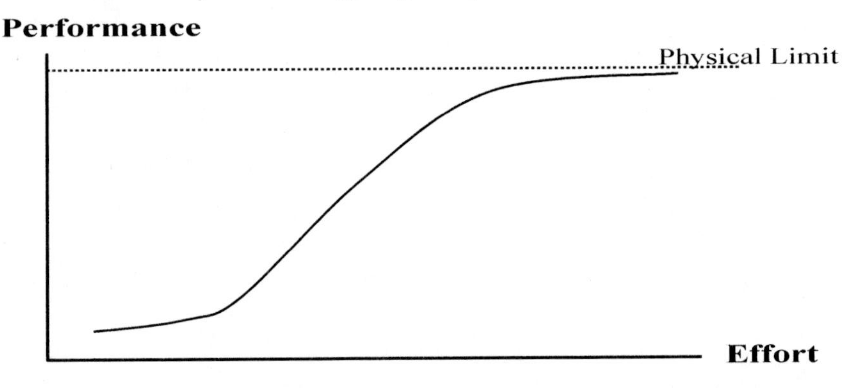

It is worth noting that the technology trajectory can apply to either product or process. Further, the product can result in the spin-off of many related products based on innovation or market segmentation, such as the Sony Walkman or a pharmaceutical due to the identification of beneficial side effects.[23] As the product is adapted to these alternative markets, it may improve in general and/or from the perspective of the alternative market(s). Alternatively, as a process technology improves, it typically results in the creation of entirely new products never seen before, a better way to make existing products not previously made with this process, and a better way to make products that were previously being made with the same process technology.[24] Prahalad and Hamel (1994) describe the power that unique knowledge and production skills can offer a firm by enabling it to be not only the market leader for some product in some industry, but to dominate multiple product markets in different industries by exploiting leadership with a product or process technology that is relevant across many industries and maintaining competitive leadership as firms move along the technology trajectory (see Figure 2).

The possibility of progressing along a given technological trajectory or jumping from one technological trajectory to another (see Figure 3) has led to a discussion of what is the most suitable strategy. This has been described in a variety of different ways. Progression along a technological trajectory is often associated with the rise of Japanese industry,[25] while others such as Christensen (1997) speak to the movement from one technological trajectory to another. This unsurprisingly is an active area of discussion and research and should continue to be so for some time. Having considered how the dynamics of technology relate to the product perspective, we now focus on innovation from the perspective of the product. While the product clearly relies on technology for its invention and development, the product perspective and associated literature is clearly different.

Figure 3. Presence of multiple different technological trajectories

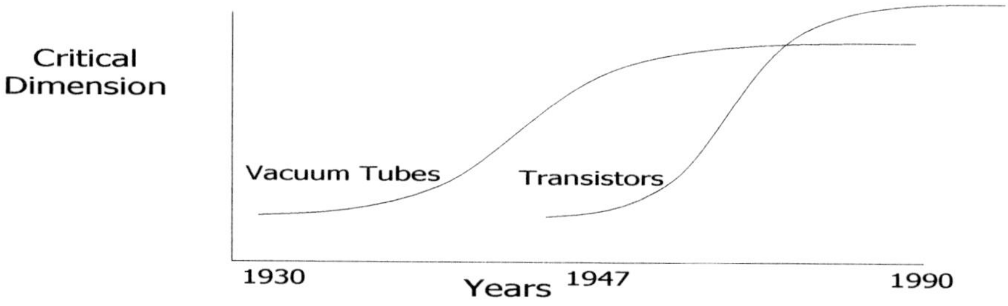

Critical Dimension

Vacuum Tubes

Transistors

1930 Years 1947 1990

THE PRODUCT PERSPECTIVE IN INNOVATION

There is a rich and dense literature that considers the product as a unit or level of analysis in the study of innovation. It is worth restating here that the intention of this chapter is to focus on contributions in books, since the seminal journal-based literature is discussed throughout subsequent chapters. Consequently, some readers are likely to ask, *But why has this important model or finding been left out?* The answer is that these results are reported in journal articles and considered elsewhere.

To start with, we consider the different stages through which a product goes without taking into account modification and innovation within a product, as discussed in the last section. Such a model must consider both the development of the product and the process in which the product ends up being used to produce the intended benefits for one or more product users. In some cases, this model is shown as one diagram and referred to as the dual model of innovation (Tornatzky & Fleischer, 1990). In other cases, the steps involving the development of a product are referred to as the *product-based model,* while the steps that involve the transfer of the product

Figure 4. Dual-based model of innovation

BASIC RESEARCH
⇓
APPLIED RESEARCH
⇓
DEVELOPMENT
⇓
TESTING/EVALUATION
⇓
MANUFACTURING
⇓
MARKETING/DISSEMINATION

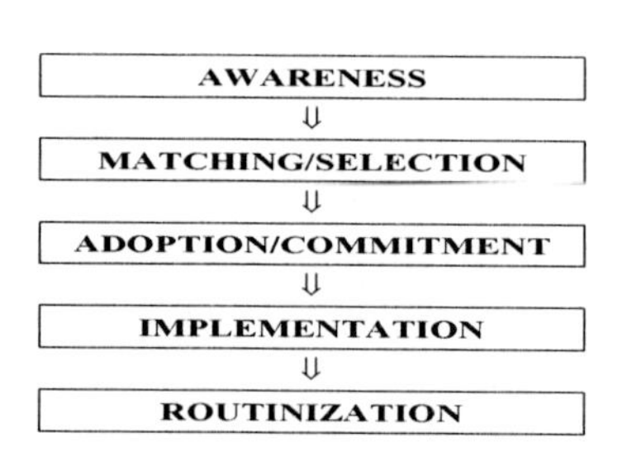

to regular application by the user are referred to as comprising the *process-based model of innovation.* The dual model of innovation and two sub-models of which it is composed are shown in Figure 4. An in-depth discussion of each of these steps can be found in many texts and is beyond the scope of this chapter. Worthwhile sources for further information on invention and the product development process include: Clark and Wheelwright (1992), Cooper (2001), and Kahn (2004). Sources of information for the *process-based model* include Tornatzky and Fleischer (1990), Rogers (1995), and Yin's (1978) work on routinization of innovation.

As we know from our earlier consideration of the work of Foster (1986), over time there can be tremendous changes to a product and the range of benefits the product offers as a result of advances along a technological trajectory. However, Foster perceives the change from the perspective of a technology. Utterback (1994) makes a critical contribution by considering the dynamics of innovation from the perspective of a product. He describes how a typical product first undergoes increasing rates of product innovation (see Figure 5). Through changes to a product, it is possible to improve its utility. At some point, however, the rate of product innovation declines. These declines may result from the product reaching its optimal form or from the product being sufficiently close to the end of the product lifecycle so that changes to design are no longer warranted. Process innovation initially has a much lower intensity than product innovation; however, as experience increases with the manufacture of product and the form of the product becomes more stable—usually associated with declining levels of product innovation—the intensity of process

Figure 5. Utterback's model of product and process innovation

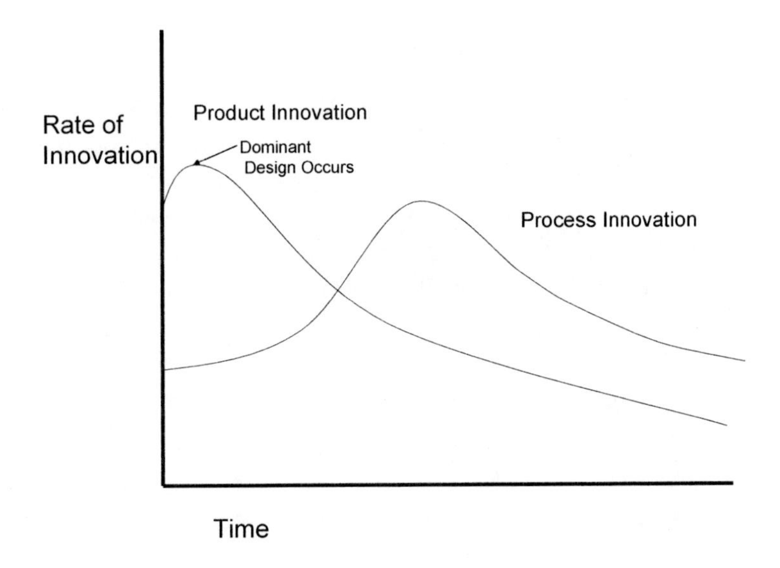

innovation increases. Process innovation is often driven by attempts to reduce the cost of manufacture, but also offers improvements in quality and consistency of a product. Finally as a product reaches maturity and sales start to decline, the rate of process innovation for the product declines.[26]

There is much book-based literature relating to product development and bringing these products to market. However, this material tends to be advice to practitioners and entrepreneurs, as opposed to the theoretical treatments that we are focused on. For example, Cooper, Edgett, and Klienschmidt (1998) offer a detailed examination of selection and evaluation of R&D at large firms. The text gives excellent examples of the different ways in which firms evaluate products and R&D. But it clearly is focused on practice. For many academics, however, such books are important since they assist the researcher in establishing benchmarks of practice and to better ensure that their theoretical work is linked to, grounded in, and applicable to practice.

The bulk of the academic research relating to the dual innovation process is on the diffusion stage of this process. Thousands of papers have been written on the diffusion stage, due to its applicability to such a wide range of fields.[27] Fortunately, Rogers (1995) has distilled this large body of work to something much more manageable. Diffusion considers the spread, or failure to spread, of an innovation through the potential-user population. Diffusion research began in the 1940s with rural sociologists studing the spread of hybrid seed. Typically, the diffusion of innovation takes the form of an S-curve (see Figure 6). Empirical work has identified great similarities between adopters on different parts of this curve. Members of a population are described by five different categories:

1. **Innovators:** These adopters are pioneers that like to be the first to try out an innovation. They are socially outside the mainstream and might even be considered eccentric. They comprise the first 2.5% of the population to adopt an innovation.
2. **Early adopters:** These adopters command respect in their communities. Their adoption of an innovation is taken by many that the innovation is worthwhile. Once early adopters signal the population that an innovation is ready for use, the rate of adoption in the population typically accelerates. They comprise the next 13.5% of the population.
3. **Early majority:** These adopters get involved with an innovation once it is clear that it has been "broken in." Frequently, this group obtains the most benefit from the innovation, since the innovation is ready for use and their adoption is early enough to extract benefit before half of the population or more adopts. They comprise the next 34% of the population.
4. **Late majority:** These adopters only adopt innovations that are widely accepted. The focus of this type of adopter is on avoiding mistakes. Consequently, an

innovation is only adopted once no risk is seen as remaining. They comprise the next 34% of the population.

5. **Laggards:** These adopters are very conservative and only make changes when they are forced to. They comprise the remaining 16% of the population.

In addition to placing adopters into different types of groups and characterizing the members of these groups in detail, the innovation literature has identified a number of characteristics that affect the rate of diffusion.

1. **Structure of communication networks:** The greater the communication between members of a population, the faster the rate of diffusion.
2. **Standardization:** If an innovation becomes the standard with a population or sub-population, it will diffuse much faster.
3. **Complexity:** The rate of diffusion is affected by its *perceived* complexity. Consequently, if an innovation is not complex, but is *perceived* as complex, it will diffuse slowly, while an innovation that is *perceived* as not complex will diffuse faster regardless of its complexity.
4. **Visibility:** The rate of diffusion is positively affected by the demonstration of its benefits. If a potential user can *see* the benefits than an innovation offers, they are more likely to adopt an innovation. The more visible the benefits are, the more rapid the rate of innovation typically is.
5. **Divisibility:** Innovations that can be adopted in an incremental fashion are adopted more frequently and rapidly. By only needing to adopt a small quantity of an innovation, risks to a user or a user's organization are minimized. By adopting only a small amount of the innovation, the user is able to evaluate an innovation, assess benefits, and gain competence in using and extracting benefits from the innovation, while minimizing the downside risk to the organization.
6. **Testability:** This is closely related to divisibility. The easier it is to test the benefits associated with an innovation, the more rapidly the innovation will diffuse through a potential user population. An innovation that is easy to test allows for the assessment of benefits with an expenditure of little time, cost, and downside risk. Innovations that are easy to test tend to be adopted on a trial basis and then adopted permanently or broadly if the test results are favorable, while innovations that are difficult to test tend to diffuse slowly or sometimes fail to diffuse through the population.

While diffusion is quite visible and as such is easy to study, the steps that follow—implementation and routinization—are critical to the practitioner but much

Figure 6. Typical diffusion of an innovation through a potential user population

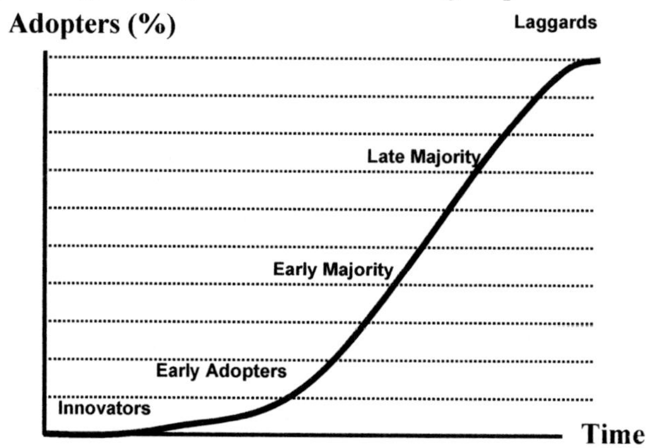

more difficult to study. Work in these areas, whether book or journal article, tends to be practice oriented.

Having considered the innovation process, it may appear that the product perspective has been addressed. However, consideration of the product perspective would be incomplete without looking at the initially unanticipated effects that the customer or user may have on a product. Unanticipated effects and uses of technology (and products) were referred to already, but the study and insights associated with lead users have yet to be mentioned. von Hippel (1988) considers the user as a source of innovation. Only mention is made of the role of users as sources of innovation, since this stream of research is also considered later on in the text. From the perspective of the product, we now move to consideration of innovation from the perspective of a market or industry.

INDUSTRY OR MARKET PERSPECTIVE

The industry or market perspective builds on both technology and product. Typically markets and industries will continue with a series of products and processes until superior products and/or processes force firms either to change or disappear (see Figure 7). Examples include the manufacture of glass using the float process and the replacement of gas lighting with electricity.[28] From this vantage point, the relevance of Foster's (1986) works on technological trajectories is important. Consideration of techniques such as roadmapping[29] and the formation of pre-competitive alliances[30] are important in practice, but are outside of the scope of this chapter due to their reliance on academic papers and their focus on practice rather than theory. The retreading of old ground allows us to offer a rather terse consideration of innovation

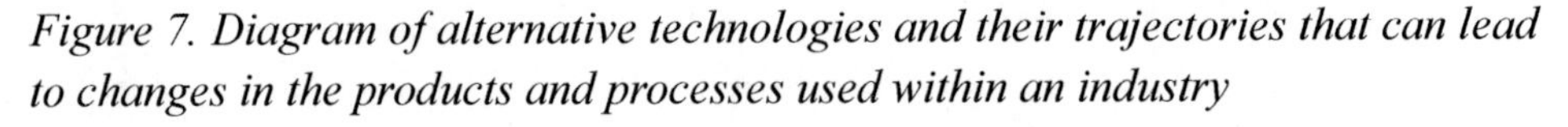

Figure 7. Diagram of alternative technologies and their trajectories that can lead to changes in the products and processes used within an industry

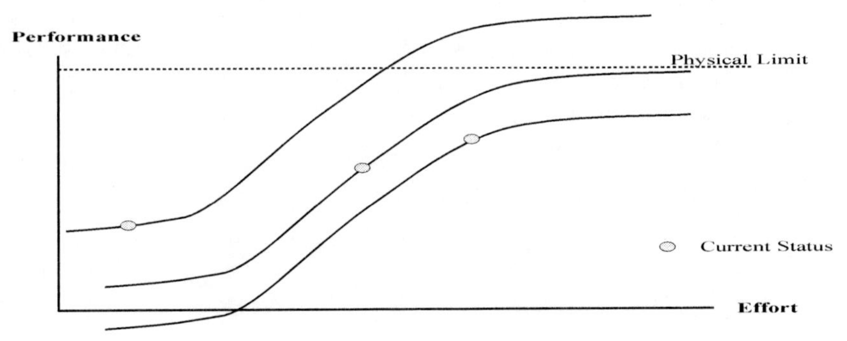

in a given industry at an even more macro level than an industry. We now consider innovation from a geographical viewpoint.

GEOGRAPHIC PERSPECTIVE ON INNOVATION

The regional or national perspective on innovation draws artificial geographic boundaries around an innovation. Some of the issues associated with geography have already been considered in our discussion of diffusion earlier in the chapter. The contributions specific to this geographic perspective typically relate to either the benefit obtained by close proximity—whether this be firms within a science park or related companies within a specific geographic regions[31]—or through the provision of the *correct environment* for innovations to occur. Consideration of success factors tend to be focused more on practice than theory. It is not clear whether this practical focus is driven by the great rewards that could result in a region if geographic density of certain types of business is either truly a critical factor for economic prosperity, or if the marketing function of discussing clusters of firms is helpful in attracting new business to specific regions.

The other part of the geographic perspective relates to the effect of policies, legislation, and spending that encourages or discourages innovation. This literature seeks to understand opportunities for taking the different actors within a geographic area and creating synergies between universities, private and public research organizations, for-profit organizations, government agencies, and non-governmental organizations to create an overall output that is greater than the sum of the individual parts. Relative wealth, population size, past history, and economic structure all are found to be important variables (Nelson, 1993).[32] While the material offered to consider this perspective is relatively brief, it is not a reflection on the importance and potential contribution of this perspective. Methodological concerns such as the difficulty in obtaining a sufficient sample size and the richness of contextual variables complicating statistical analysis clearly creates barriers to study and development of generalizable theory. The geographic perspective is clearly an area in which one expects to see a tremendous amount of effort in the future. Having considered the geographical perspective, which in many cases is a socially constructed and artificial creation—a country—we turn our attention to the other critically important socially constructed creation—the company.

ORGANIZATIONAL PERSPECTIVE ON INNOVATION

There has been much written and great consideration given to the organizational perspective on innovation. The focus of many books has been on organizational structure or innovation's impact on organizations. Books considering organizational structure have provided views on how different ways of organizing a firm have had different effects on innovation. Burns and Stalker (1966) considered how firms that had a more flexible organic structure were more effective at working with innovation. However, firms with a more mechanistic structure were more effective once the innovation was integrated into the firm. Many people have suggested the placing of a 'firm inside a firm' to isolate the innovative from more mundane day-to-day routines. An excellent account of this technique is Kidder's (1981) consideration of the development of a new computer. Further insights into the question of overall structure are addressed by Morgan (1986), who not only considers the mechanistic and organic structure, but many other *lenses* through which an organization can be *seen*. Others such as Kanter (1983) consider the development of an environment that encourages innovation from more of a humanistic and leadership perspective.

The other theme is how innovation affects the firm. Innovation is discussed both from the perspective of threat and opportunity (Foster, 1986; Christensen, 1997). Innovation is a threat since it can change the competitive landscape and opportunities of an organization's environment. For the same reasons innovation also offers an opportunity. Rogers (1995) considers how firms do or do not adopt these innovations, while Iansitti (1997) considers how firms integrate these adoptions into their organizations. Having considered a single unit, the organization, we now move to two or more organizations working or interacting with each other—the network.

NETWORK PERSPECTIVE ON INNOVATION

When one mentions the network perspective and innovation, the book that researchers tend to refer to is Tom Allen's (1977) work on the flow of information within research organizations. This study on informal work networks within a laboratory highlights how the pathways of communication within a laboratory environment are quite different from the formal hierarchical structure of the laboratory. Undoubtedly, this contribution was critical in attracting psychologists and sociologists to study how the management of scientists and technologists is different from other types of employees. Allen discusses the importance of individuals that span the boundaries of the organization either through contact with people in other organizations or through reading the literature produced outside the organization. Not only is this work a first for its identification of the differences in the working behavior and consequently the way in which scientists should be managed, but it also is indicative of the need to consider relationships between internal members of an organization and outsiders that are not formally recognized by the participating employee organizations—that is, relations outside of joint ventures or other formal agreements. Having recognized the value of this work, it is worthwhile to point out the different types of networks.

As indicated earlier, there are a variety of different networks.[33] Due to the focus on organizations in the study of management, its researchers often consider networks to be a network of organizations working together through a series of formal relationships.[34] Alternatively, networks can involve the consideration of actors within an organization, as in the communications networks between the scientists in the research setting studied by Allen. Or a network can extend outside of the organization, indicated by Allen's work and pursued subsequently by other researchers. Whether networks are internal or external, or being measured at the firm or individual level, it is important to recognize that networks can involve formal or informal relationships. In summary, the following types of networks are likely to be encountered in the research on management of innovation and technology:

- Formal networks of organizations,
- Formal networks of people within an organization,
- Formal networks of people extending across organizations,
- Informal networks of people within an organization,
- Informal networks of people extending across organizations, and
- Informal networks of organizations.

The network perspective will be considered further, both in the chapters that follow and the final chapter that considers the direction and trends of TIM research. Finally, innovation can be considered from the perspective of the individual.

INDIVIDUAL PERSPECTIVE ON INNOVATION

The individual perspective on innovation considers innovation's development, application, and use. Many people think of the lone inventor working tirelessly until one comes across the grand discovery, which is followed by shouts of *Eureka* and much excitement. The individual perspective was found by Allen (1977) to involve scientists playing different roles in a laboratory—most notably that of the boundary spanner. The boundary spanner interacts with the external environment either through reading the literature or through interaction with individuals and/or organizations outside of the laboratory. The boundary spanner is an important role since he or she connects members of the organization to information sources that would otherwise be inaccessible. Other roles have been identified in the literature, as we will see in the chapters that follow. Additional consideration is given in various places on how to encourage people to be more innovative or better identify and select innovative people.

The application of innovation is an important area involving the individual perspective. von Hippel (1988) recognizes individuals as being important in identifying new uses or determining innovative improvements to existing products. von Hippel terms these individuals as 'lead users'. Since a number of his papers are considered further in this book, discussion is limited to the mention of his main ideas in his important book, *The Sources of Innovation.* Finally, the use of innovation is considered. The diffusion literature, once again, is found to be important (Rogers, 1995). The characteristics associated with those who adopt innovation at different times are considered in depth by this field. While considering the use of innovation, it is important to recall the concerns and opportunities offered by unanticipated consequences of innovation. See not only the horrific stories relating to innovation and technological change alluded to near the start of the chapters, but the opportunities that are identified by *lead users.*

The user perspective has taken on new meaning over the last 20 years with the rise of management information systems (MIS). The adoption of software systems across large numbers of people, within and across organizations, has led to increased interest in the use of innovation. Due to the relative youth of this field, the user perspective is best dealt with through the consideration of articles, not books. A final note: whether one is using an MIS innovation or an innovation in some other field, it is important to consider how continuous or discontinuous an innovation is.

A continuous innovation is one that the user does not have to change their behavior at all to engage—for example, a piece of software that has a familiar interface but does different calculations in the background. An innovation that is discontinuous requires different behavior from the user. Unsurprisingly, as discontinuity increases, it is more difficult to have the innovation either used or adopted. Having considered innovation from a variety of different perspectives, an explanation of the methodology used to select the papers and topics considered in Chapters II through IX is now offered, followed by some concluding notes.

THE STRUCTURE OF THE REST OF THIS BOOK: THE LOGIC AND METHODOLOGY EMPLOYED

This book is the end product of an assignment started by Desiree Roberts as a doctoral student studying under the direction of Jonathan Linton. The initial thought or concern was how *every academic she met would have read articles and have knowledge she lacked.* But what was important was to be familiar with everything that one should know; sadly, in too many cases people are not. In order to achieve this goal in the discipline of technology and innovation management (sometimes referred to as TIM or MoT), Jonathan Linton and George Biggar of the RPI research library[35] put together a collection of the most frequently cited articles related to technology and innovation management that could be found on the *Web of Science.* Tables in the book show the citation frequency of each article in 2004 (when the initial independent study course occurred) and 2007 (as we finalized this book). These tables are interesting because they offer insight into what articles and topics, while important in the past, have already matured to the point that there is not much written over the last few years, as opposed to other articles that have recently experienced an intensification of interest and effort with questions related to this seminal work. This interest is reflected in a substantial increase in the number of times a paper was cited between 2004 and 2007.

Dr. Roberts made excellent and extensive notes on each paper that was on the list of seminal technology innovation management papers. The result: a substantial amount of information that was exceedingly useful in offering a foundation in technology and innovation management. However, the density of the information and the fact that the relationship between one article and the next was the alphabetical order of the author's name made the reading of this work somewhat less than pleasurable. While there are many good books written on specific topics in the Technology and Innovation Management area, future directions of research, and collections of interesting readings, there is a gap in the consideration of the foundational literature in technology and innovation management.[36] Consequently, we took up IGI Global on their interest in and offer to publish a book that pursued this objective. First, the paper summaries were grouped into themes. This was done through the use of the content expertise of Robert Friedman, and at the same time independently while using a self-organizing map (discussed further in the final chapter). The results of both of these sorting mechanisms were found to be quite similar. Jonathan Linton then used his content expertise to look at the overall organization and cases in which a paper appears to better belong in one section rather than another. In cases where it was recommended that papers be moved from one chapter to another, Drs. Friedman and Linton discussed each individual case and came to a mutual agreement on the best location for the chapter. For example, some of the papers that relate to management information systems initially appeared in other chapters, but a decision was made to group all MIS articles together.[37]

This introductory chapter has considered the different perspectives that currently exist in the TIM/MoT field, and in the process has reviewed the seminal ideas that have appeared in books, written by Jonathan Linton. The earlier mentioned alphabetized notes of Dr. Roberts were transformed into Chapters II through IX by Dr. Friedman, who also included several articles noted by early reviewers. As mentioned previously, Chapter X was added specifically by Dr. Friedman to satisfy the needs of the students of Information Technology. Finally, concluding parts of Chapters II through IX and the final chapter were written by Dr. Linton. We all, however, stand by the entire book as being as accurate, complete, and engaging as we could make it; having offered an explanation of how the book came about and some insights into the logic of its flow, concluding notes are offered for this introductory chapter.

CONCLUSION

Whether you are interested in Innovation Management, the management of technology (MoT), or technology innovation management (TIM), it is important to understand that this is a complex field of study for a variety of reasons:

1. The study of business and management often focuses on or assumes a static or equilibrium condition. By its very nature, innovation is dynamic. Consequently, assumptions of stasis that are theoretically convenient and/or offer tractability and closed-form solutions are fundamentally flawed.
2. Innovation can be considered from many different perspectives. The different perspectives offer different insights into this field. While it is not necessary to study or to contribute to all of these different perspectives, it is critical to recognize that many different perspectives exist, that none of these perspec-

tives are incorrect, and that the denial of the existence or value of different perspectives is problematic and will lead to lost opportunity and limited understanding. Having said this, individual researchers and practitioners do not need to pursue all perspectives, but need awareness of their existence. Our consideration of a number of different perspectives and recognition of the important contributions made by each has hopefully made this clear. This point cannot be overstressed, since it is common to see academics cast aspersions on other research due to the use of alternative perspectives, units, and levels of analysis, if not directly, then indirectly through an attack on the data sources and methods that are commonly used to consider such research. For example, a large *n* empirical study is a logical approach if innovation is being considered from the perspective of the individual. However, such a study is not possible when considering issues related to national policy—a perspective with small populations available for study:

3. The innovation field is inter- and multi-disciplinary by nature. Innovation can be studied using the theories, methodologies, and assumptions of a variety of different fields. This creates a difficulty in terms of communication, since different communities of practice are not accustomed to communication with each other. The study of innovation is represented in many fields, journals, and organizations. In some cases the overlap between these communities is limited at best. For example, a typical innovation researcher may attend two or three conferences per year. If one were to attend *all* the important innovation management conferences in a given year, one would be attending a large number, including but not limited to: IAMOT, IEEE EMS, PDMA, PICMET, and R and D Management.[38] In addition, there are a number of conferences one might attend that are more general in focus, such as: AOM, DSI, and INFORMS.[39] Finally, it is likely that a researcher could easily justify attending one or more local or domestic conferences that are topical. This is in contrast to the number of events in a traditional discipline, where attendance at two to three conferences is sufficient to keep track of developments in one's field. The inter-disciplinary nature presents challenges and opportunities. Consequently, this field is different from many traditional fields of study.
4. A related but different issue worth noting is that many different fields of academic study offer valuable insights to the field of innovation. The different perspectives in some cases span many of these fields, but rarely span all of them. These fields include:
   - *Psychology,* and allied fields of organizational behavior, human resources, and marketing;
   - *Economics,* and allied fields of accounting and finance;
   - *Information sciences,* and allied fields of management information systems and information technology;

- *Management science,* and allied fields of operations management, operations research, and industrial engineering;
- *Sociology,* and allied fields of organization behavior, human resources, communication, marketing, policy, and strategy;
- *Engineering,* and allied fields of design, manufacturing and industrial engineering, and operations research and management;
- *Political science,* and the allied field of policy; and
- *History,* particularly the history of economics, and of science and technology.

5. Many ideas are relevant to different perspectives. Consequently, some researchers and practitioners may have in-depth knowledge of certain theories and ideas that appear to be from one perspective or traditional discipline, but are unaware of other equally important contributions. When this situation occurs it is often because the idea or theory under consideration spans multiple areas. For example, Rogers' (1995) work on diffusion is widely recognized and known by innovation researchers. Diffusion can be considered from a number of different perspectives: individual, network, organizational, geographic, and technology. Also, diffusion resides as a topic of study in a number of different traditional disciplines, including economics, journalism, sociology, political science, and psychology. Even with this very broad acceptance and awareness, competing approaches and definitions can be offered. In the case of diffusion, there is the user-based study and definition offered by Rogers (1995), as opposed to the technology-based definition offered by Kuznets (1979).

For those interested in innovation, at least some awareness of the different perspectives is required, and the first chapter of this text lays the groundwork for this goal. Through the consideration of the seminal, most frequently cited journal literature in the chapters that follow, we provide the structure and information for the achievement of this goal.

REFERENCES

Allen, T.J. (1977). *Managing the flow of technology.* Cambridge, MA: MIT Press.

Argote, L., & Epple D. (1990). Learning curves in manufacturing. *Science, 247*(4945), 920-924.

Argyris, C., & Schon, D. (1978). *Organizational change.* Reading, MA: Addison-Wesley.

Barras, R. (1986). Towards a theory of innovation in services. *Research Policy, 15,* 161-173.

Bohn, R.E. (1994). Measuring and managing technological knowledge. *Sloan Management Review, 3,* 61-73.

Beyer, D.E. (1991). *The Manhattan Project: America makes the first atomic bomb.* New York: Watts.

Beyer, J.M., & Trice, H.M. (1978). *Implementing change.* New York: The Free Press.

Burns, T., & Stalker, G. (1966). *The management of innovation.* London: Tavistock.

Cameron, J. (1984). *The terminator* (film).

Carson, R. (1962). *Silent spring.* Boston: Houghton Mifflin.

Chaplin, C. (1936). *Modern times* (film).

Christensen, C. (1997). *The innovator's dilemma: When new technologies cause great firms to fail.* Boston: Harvard Business School Press.

Cooke, P., & Morgan, K. (1998). *The associational economy: Firms, regions, and innovation.* New York: Oxford University Press.

Cooper, R.G. (2001). *Winning at new products: Accelerating the process from idea to launch.* New York: Perseus Books.

Cooper, R.G., Edgett, S.J, & Kleinschmidt, E.J. (1998). *Portfolio management for new products.* Reading, MA: Perseus Books.

Crum, W.L. (1923). Cycle rates of commercial papers. *Review of Economic Statistics, 5*(1), 17-29.

Devezas, T.C., Linstone, H.A., & Santos, H.J.S. (2005). The growth dynamics of the Internet and the long wave theory. *Technological Forecasting and Social Change, 72*(8), 913-935.

Feldman, M. (1994). *The geography of innovation.* Boston: Kluwer Academic.

Foster, R.N. (1986). *Innovation: The attacker's advantage.* New York: McKinsey and Company.

Garcia, M.L. (1997). *Fundamentals of technology roadmapping.* Albuquerque: Sandia Laboratories.

Hakansson, H. (1989). *Corporate technological behaviour: Cooperation and networks*. London: Routledge.

Hightower, J. (1972). *Hard tomatoes, hard times: The failure of America's land-grant college complex.* Cambridge, MA: Schenkman.

Hung, S.C., & Chu, Y-Y. (2006). Stimulating new industries from emerging technologies: Challenges for the public sector. *Technovation, 26*(1), 104-110.

Iansitti, M. (1997). *Technology integration: Making critical choices in a dynamic world.* Cambridge, MA: Harvard Business School Press.

Ishikawa, K. (1989). *Introduction to quality control.* White Plains, NY: Quality Resources.

Jonnes, J. (2003). *Empires of light: Edison, Tesla, Westinghouse, and the race to electrify the world.* New York: Random House.

Juglar, C. (1967) [1860]. *Des crises commerciales et leur retour périodique en France, en Angleterre, et aux États-Unis*. New York: A.M. Kelley.

Kahn, K.B. (2004). *The PDMA handbook of new product development.* New York: John Wiley & Sons.

Kanter, R.M. (1983). *The change masters: Innovations for productivity in the American corporation.* New York: Simon & Schuster.

Kishi, T., & Bando, Y. (2004). Status and trends of nanotechnology R&D in Japan. *Nature Materials, 3*(3), 129-131.

Kidder, T. (1981). *The soul of a new machine*. Boston: Little Brown.

Kitchin, J. (1923). Cycles and trends in economic factors. *Review of Economic Statistics, 5*(1), 10-16.

Kondratieff, N.D. (1984). *The long wave cycle*. New York: Richardson & Snyder.

Kurlansky, M. (2002). *Salt: A world history.* New York: Walker.

Kuznets, S.S. (1979). *Growth, population, and income distribution: Selected essays*. New York: Norton.

Lane, N., & Kalil, T. (2005). The national nanotechnology initiative: Present at the creation. *Issues in Science and Technology, 21*(4), 49-54.

Liker, J.K., Ettlie, E., & Campbell, J.C. (1995). *Engineered in Japan: Japanese technology-management practices*. New York: Oxford University Press.

Linton, J. (1999). Developing skills with existing technology. *Circuits Assembly, 10*(1), 26-28.

Linton, J.D., & Walsh, S.T. (2004). Integrating innovation and learning curve theory: An enabler for moving nanotechnologies and other emerging process technologies into production. *R&D Management, 34*(5), 513-522.

Lundvall, B.A. (1992). *National systems of innovation: Towards a theory of innovation and interactive learning.* London: Pinter.

Meadows, D.L. (1992). *The limits to growth; a report for the Club of Rome's project on the predicament of mankind.* New York: Universe Books.

Morgan, G. (1986). *Images of organization*. Beverly Hills, CA: Sage.

Nelson, R. (1993). *National innovation systems: A comparative analysis.* New York: Oxford University Press.

Owen, D. (2004). *Copies in seconds: How a lone inventor and an unknown company created the biggest communication breakthrough since Gutenberg.* New York: Simon & Schuster.

Parfit, M. (2000). *Australia—a harsh awakening. National Geographic, 198*(1), 2-31.

Plato. (1987). *The republic*. New York: Penguin.

Prahalad, C.K., & Hamel, G. (1994). *Competing for the future*. Boston: Harvard Business School Press.

Reynolds, M.D. (1994). *How Pasteur changed history: The story of Louis Pasteur and the Pasteur Institute*. Bradenton, FL: McGuinn & McGuire.

Roberts, E.B. (1988). Managing invention and innovation. *Research Technology Management, 1,* 11-29.

Rogers, E.M. (1995). *The diffusion of innovations* (4th ed.). New York: The Free Press.

Sanderson, S.W., & Uzumeri, M. (1996). *The innovation imperative: Strategies for managing product models and families*. New York: McGraw-Hill.

Saxenian, A.L. (1994). *Regional advantage: Culture and competition in Silicon Valley and Route 128*. Cambridge, MA: Harvard University Press.

Schnaars, S.P., Swee, L.C., & Maloles, C.M. (1993). Five modern lessons from a 55-year-old technological forecast. *Journal of Product Innovation Management, 10,* 66-74.

Schumpeter, J.A. (1934). *The theory of economic development.* Cambridge, MA: Harvard University Press.

Schumpeter, J.A. (1939a). *Business cycles: Volume 1.* New York: McGraw-Hill.

Schumpeter, J.A. (1939b). *Business cycles: Volume 2.* New York: McGraw Hill.

Schumpeter, J.A. (1952). *Capitalism, socialism and democracy.* New York: Harper and Row.

Smith, A. (2005) [1776]. *An inquiry into the nature and causes of the wealth of nations.* London: Routledge.

Stokes, D.E. (1997). *Pasteur's quadrant: Basic science and technological innovation.* Washington, DC: Brookings Institution Press.

The Economist. (1999). Catch the wave. *The Economist, 350*(8107), 7-8.

Tornatzky, L.G., & Fleischer M. (1990). *Processes of technological innovation.* Lanham, MD: Lexington Books.

Utterback, J.M. (1994). *Mastering the dynamics of innovation: How companies can seize opportunities in the face of technological change.* Boston: Harvard Business School Press.

von Hippel, E. (1988). *The sources of innovation.* New York: Oxford University Press.

Wachowski, A., & Wachowski, L. (1989). *The matrix* (film).

Wasserman, S., & Faust, K. (1994). *Social network analysis.* New York: Cambridge University Press.

White, L. Jr. (1962). *Medieval technology and social change.* New York: Oxford University Press.

Winickoff, D.E. (2006). Governing stem cell research in California and the USA: Towards a social infrastructure. *Trends in Biotechnology,* 24(9), 390-394.

Yin, R.K. (1978). *Changing urban bureaucracies: How new practices become routinized.* Santa Monica, CA: Rand Corporation.

ENDNOTES

[1] A discussion of the development of electric lighting can be found in Utterback (1994) and Jonnes (2003). Also Thomas Edison's laboratory in West Orange,

New Jersey, is well worth a visit to those interested in the history of technology and new product development.

[2] The term *applied benefits* is used to suggest a benefit other than an increase in knowledge. These benefits may directly result in one or more new products or result in the ability to improve one or more performance characteristics of a product. Alternatively, the benefits might be more indirect, such as resulting in improvements in a process that allows for the improvement of existing products or the development of new products.

[3] See for example Reynolds (1994) for further insights into the work of Pasteur.

[4] This definition is very close to Rogers' (1995). Other definitions are more specific, such as innovation = invention + exploitation (Roberts, 1988).

[5] Change management falls outside the scope of this book. But there are many good sources that will offer insights into this area, including Argyris and Schon (1978) and Beyer and Trice (1978).

[6] The reader interested in pursuing this line of thought further will find Bohn (1994) to be of interest and helpful.

[7] In Hightower's study of the automated tomato picker, the unintended consequences resulting from the introduction of a mechanical tomato picker that was intended to help farmers are chronicled. The consequences of the tomato picker include: the change in taste and texture of tomatoes, the destruction of family farms, and the dislocation of migrant farm workers.

[8] This study is a mixture of striking color pictures and text explaining the unanticipated consequences and environmental devastation resulting from the intentional introduction of non-native species to Australia.

[9] See previous discussions of the problems and solutions associated with ozone depletion or more recent consideration of global warming.

[10] Meadows (1972), in the book *The Limits to Growth,* uses the techniques of systems analysis and simulation to make very dire predictions about the Earth's future based on changes of a number of key variables such as population, food supply (and associated technology), raw materials, and pollution. The predictions have fortunately been incorrect. Schnaars, Swee, and Maloles (1993) consider a forecast made decades ago regarding the future impact of technologies, such as recreational vehicles and the fax machine. The errors in forecasting are evident decades later and make for interesting reading.

[11] Kondratieff's location was not only unfortunate for him, but also for the diffusion of his work and thought. The diffusion process of the work was slowed down by the hostility of the local political system to it, but also to its publication in Russian in 1928. It was not until 1935 that a translation abridged from

the German work translation of 1926 was published, and in 1984 the original text was translated from Russian directly into English and published.

[12] The omissions of guidance in terms of the type of technology or degree of maturity required for the onset of a new Kondratieff wave does open his theory up to charges of being un-testable.

[13] Either explanation is possible since the threat of censure was very clear in the Soviet Union—he was in fact sent to a gulag. From a scientific perspective, Kondratieff may have been dissatisfied with his ability to determine the initiation date and point at which the technologies diffused sufficiently to have a significant effect on economies. It is after all difficult to state the start and diffusion date of major technological advances such as the steam engine, cotton gin, and steel production that appear to be the basis of long economic waves.

[14] For a more recent perspective on Kondratieff waves, the reader is referred to Devezas, Linstone, and Santos (2005).

[15] The words *carry out* are italicized to indicate that this is not the choice of language by the authors, but the manner in which Schumpeter refers to entrepreneurs.

[16] Named after Clement Juglar (1860) and his work on the timing of economic prosperity and crises.

[17] There are many other proposed cycles of short and medium duration, but they are macroeconomic in nature and clearly outside the scope of this work.

[18] The other 25% results from factors of production such as raw materials and size of workforce.

[19] These early categories or definitions are important and are often overlooked and not cited in the later literature. Many different terms have been used recently. The term *fundamental innovations* seems to relate at least somewhat to terms such as disruptive, radical, major, and revolutionary. For innovations that offer additional or small improvements over the initial *fundamental* innovation, relevant currently used terms include: normal, incremental, routine, sustaining, and evolutionary. The challenge with dealing with this terminology is the tendency for the terms to be used interchangeably, even though the terms have sometimes subtly different meaning. The other problem is the use of multiple similar, but significantly different definitions for these terms. Much of the difference in definitions can be attributed to a different unit of analysis being used—for example, firm, product, or user.

[20] It is unsurprising if while reading this you think of Prahalad and Hamel's (1994) *Competing for the Future* or Christensen's (1997) *The Innovator's Dilemma.*

[21] See also Linton (1999).

[22] The learning curve literature is quite extensive. Reviews of this literature can be found in Linton and Walsh (2004) and Argote and Epple (1990).

[23] This can happen in different ways, for example: (1) new uses that are unapproved as discovered by physicians are adopted by other physicians—referred to as off-label use, and (2) alternative applications of a medication are discovered during the appearance of useful side effects. Viagra was developed as the result of the *useful* side effect of a medication being identified and developed for a different purpose.

[24] These three items are placed in reverse order of temporal occurrence. Typically, as one does a better job producing the same product, one discovers that they can now produce other products, and finally one discovers that it is possible to produce an entirely new product that did not exist previously (see Barras, 1986).

[25] See: Liker, Ettlie, and Campbell (1995) or Ishikawa (1989).

[26] With the prospect of an ever-declining number of total sales, fewer and fewer innovations become economically attractive to pursue.

[27] Research is in a wide range of fields including agriculture, engineering, management, journalism, marketing, medicine, policy, political science, and sociology.

[28] See Utterback (1984).

[29] For further consideration of roadmaps, see Garcia (1997).

[30] Examples include Esprit and Sematech for the semiconductor industry and ICOLP for the elimination of ozone-depleting chemicals from electronics manufacturing firms.

[31] For example, Saxenian's (1994) study of the Silicon Valley and the Boston area.

[32] For further insights on how government policy and innovations systems vary between countries, see Nelson (1990), Lundvall (1992), Feldman (1994), and Cooke and Morgan (1998).

[33] For a good overview of the network literature, the reader is directed to Wasserman and Faust (1994).

[34] For an example in the Management of Technology literature, see Hakansson (1989).

[35] A special thanks is extended here to George Biggar of the Rensselaer Polytechnic Institute research library for the tremendous support he has provided in relation to this work and in assisting Dr. Roberts with her thesis.

[36] The closest book we can find that covers the important thoughts and provides an excellent foundation is the *Diffusion of Innovations* by Everett Rogers. While we have not matched Everett for his readability, we have tried to offer similar thoroughness and value.

[37] Seminal articles in management information systems/information technology are dealt with as a separate chapter since it is the only area within Technology and Innovation Management that is currently considered a discipline on its own. In fact, MIS and/or IT is quite a unique field in that it can be found as a department within a business school, an engineering school, a computing school, or as an entity on its own in a college or university. Due to this unique position, we have included a chapter on more recent issues in management information systems and information technology. To the student of Technology and Innovation Management, this will seem an unnecessary and uncalled for diversion, in which case it is suggested that you skip the chapter. However for those who are more akin to Dr. Friedman's view, the inclusion of an additional chapter specifically addressing MIS/IT is a welcome addition that takes the seminal work of the more generalist base knowledge that is vital, but not specific to IT, and adds to it insights into IT research that would be clearly missed—if they had not been included.

[38] Information on these organizations and associated conferences is available at the following Web sites: *www.iamot.org, www.iieee.org/ems, www.pdma.org, www.picmet.org,* and *www.radma.org.*

[39] Information on these organizations and associated conferences can be found at the following Web sites: *www.aomonline.org, www.decisionsciences.org,* and *www.informs.org.*

Chapter II
R&D Process Models

INTRODUCTION

This chapter on research and development processes and models begins with a section concerning the economics and finance of R&D. Liberatore and Titus (1983) address the level and effectiveness that R&D managers have over the budgeting activities related to their projects and how best to improve these activities. Guerard, Bean, and Andrews' (1987) focus is on the financial decisions hypothesis and development of an econometric model to examine the relationships of R&D with financing decision making. Hill and Snell (1988) discuss the different stresses and influences that investors and consumers place upon the R&D process, while Hokisson and Hitt (1988) examine management structure and diversification to understand their effect on investment by external capital markets in R&D firms. Baysinger and Hoskisson (1989) are also concerned with diversification strategies that affect R&D, reporting on their empirical research findings that suggest a positive relationship between the level of R&D intensity and the level of business dominance. The section concludes with Pisano (1990) and his discussion of sources of transaction costs, particularly small-numbers-bargaining hazards and appropriability concerns, and their effect on the selection of internal or external R&D sources when technological changes affect the locus of R&D expertise.

The second section of this chapter focuses on the roles of knowledge transfer, both within R&D organizations and among various technology stakeholders. Martin and Irvine (1983) propose a model for assessing the level of contribution that different research groups of a specific content area contribute to their field's knowledge base. Gupta, Rai, and Wilemon (1986) present a discussion of how the information relationship between research and development and marketing integration is fundamental to a firm's business strategy and its approach to environmental uncertainty, while Cohen and Levinthal (1990), with the introduction of the term "absorptive capacity," focus on how individuals and the firms that employ them identify, assimilate, and use information to fuel innovation. Lane and Lubatkin (1995) refine Cohen and Levinthal's construct, absorptive capacity, by focusing on what they consider the three key elements of knowledge transfer: the knowledge offered by the "teaching" firm, the similarity between "teacher" and "student" firms regarding compensation practices and organizational structures, and the level of familiarity the student firm has with the teacher firm's organizational problems. Mansfield (1991) examines the nature and degree to which technological innovations are based on academic research and the time innovators expend in engaging academic research and industry's subsequent use of their results. Christensen and Bower (1996) investigate the relationship between technological innovative firms and their customers' demands and expectations, and why firms that are attuned to customer needs sometimes fail to produce innovations that are known to be critical to their own success. Surprisingly, too much attention to those needs results in infeasible goals and strategies.

R&D AND ECONOMICS

The Value of Financial Measurement to R&D Management

Noting the paucity of empirical studies that take quantitative techniques into account when researching R&D project management, yet acknowledging how advances in computing technologies benefit quantitative investigations, Liberatore and Titus (1983) conducted an empirical study of the use of management science techniques, selecting 29 Fortune 500 firms from a variety of industries located around the U.S. The term 'project management' as defined in this study includes "the activities of screening, selecting, evaluating, budgeting, scheduling, and controlling R&D projects." Forty respondents were interviewed, all R&D budget heads or upper management. The authors collected demographic information, data "related to familiarity and usage of project management techniques," as well as data regarding "the perceived impact of techniques on project decision-making, and any recent/planned changes in the cadre of techniques." The authors examined techniques such as "heavy use and high perceived impact of financial methods for project selection, selective use of network models, [finding] some dissatisfaction over the methods available for project scheduling and control, and no usage of mathematical programming models for R&D resource allocation" (p. 962).

While the use "of the standard measures of financial analysis for screening and evaluation of R&D projects [was nearly ubiquitous,] discounted cash flow analysis is often used selectively for those projects where the costs and rewards can be estimated with some certainty, which usually means the project has passed through the discovery phase, and is moving toward commercialization" (p. 970). In addition, "Mathematical programming models are generally not used as part of the budgeting and resource allocation process, primarily because of the diversity of project types, resources and criteria within the budgeting unit." In keeping with results from surveys from the 1960s and 1970s, the authors found "heavy usage of financial methods for project selection and evaluation, minimal usage of resource allocation models, and...increased use of Gantt charts principally for project control. Techniques and methods developed during and after the previous U.S. surveys have made some inroads into R&D project management. These include: PERT/CPM for scheduling and control, decision analysis for project evaluation, and formalized systems for budgeting" (p. 971).

Liberatore and Titus conclude that "R&D managers must have a thorough understanding of the capital budgeting techniques used by their organizations...the initial training for R&D managers in project management should provide a broad-based introduction to the available methods and techniques, while emphasizing organizational 'fit' considerations, and that "R&D managers [should] enlist the support of management scientists in the development of decision support systems for R&D project management, especially for multi-project planning and control" (p. 962).

Testing the Financial Decisions Hypothesis

Guerard et al. (1987) offer an "econometric model to analyze the interdependencies of the research and development, investment, dividend, and new debt financing decisions" in an effort to "test empirically the independence of financial decisions hypotheses." Gathering data from 140 manufacturing companies between 1978 and 1982, and citing the relevant literature, their *Management Science* article comments on "the perfect markets hypothesis and examine[s] econometrically the interdependencies of the R&D, investment, dividend, and new debt decisions." The perfect markets hypothesis suggests that "the dividend decision is independent of the investment decision by deriving that the valuation process of the firm is independent of dividend policy and firm value is dependent upon investment opportunities to produce earnings, dividends, or cash flow" (p. 1420).

After noting that earlier researchers such as McCabe (1979) have rejected the perfect markets hypothesis, the authors suggest "that firms simultaneously determine their research and development, investment, dividend, and new debt policies [which] generally are substantiated in the financial literature...Management gains insights into the interactions of pursuing research and development expenditures, paying dividends, and undertaking investments." Earlier researchers, such as Dhrymes and Kurz (1967), have concluded that there is "strong interdependence between the investment and dividend decisions; new debt issues result from increased investments and dividends but do not directly affect them; [that] the interdependence among the two-stage least squares residuals compel the use of full information (three-stage least squares regression methods); and [that] the accelerator as well as profit theory is necessary to explain investment" (Guerard et al., 1987, p. 1420).

Corporate financial officers and other fiscal decision makers are often faced with complex situations in which resources must be allocated to accomplish a variety of objectives, but those objectives may present themselves as "incompatible." Typically, a firm's managers try to manipulate "dividends, capital expenditures, and research and development activities while minimizing the reliance upon external funding to generate future profits." All of these activities can be considered to draw from "a 'pool' of resources, composed of net income, depreciation, and new debt issues," which is depleted by research and development expense, capital project expense, and dividend payments (p. 1419). The authors' econometric model, which can determine a firm's "research, dividend, investment, and new capital issue interdependence," adds credence to approaches such as Dhrymes and Kurz's, and suggests its "importance to the manager who is interested in integrating the research and development decision with the dividend, investment, and financing decisions" (p. 1426).

Competing Influences of Investors and Managers on R&D Investment

Hill and Snell (1988) present the opposition between the interests of investors and those of managers as they relate to corporate profitability. They offer a new view on how "corporate strategy might be influenced by control type, or how strategy might be a mediator in the relationship between control type and financial performance" (p. 578). Their premise is simply stated: "The divergence of interest between managers and stockholders has implications for corporate strategy and firm profitability" (p. 577). It should not be surprising that stockholders are primarily interested in seeing companies find ways to maximize their wealth, while managers are primarily interested in devising strategies that demonstrate their utility to their firm, "includ[ing] power, security, and status, as well as wealth." After examining 94 Fortune 500 companies involved in research-intensive activities, the authors conclude that "when stockholders dominate, innovation strategies are favored. When managers dominate, diversification strategies are favored. In addition, innovation is argued to be associated with greater firm profitability than diversification" (p. 577). Ultimately, corporate "governance...influences firm profitability through strategic choice" (p. 588).

Previous research has demonstrated the "divorce of ownership and control thesis," beginning with Berle and Means (1932), continuing with Marris (1964), who finds that "strategies designed to maximize firm size and diversity are compatible with strategies designed to maximize firm profitability," as well as Galbraith, Samuelson Stiles, and Merrill (1986) and Kay (1984), who find that "within research-intensive industries management will prefer a focus upon diversification, and stockholders a focus upon innovation" (Hill & Snell, 1988, pp. 577-578). It stands to reason that with stock concentration comes stockholder power, and firms "will pursue strategies consistent with maximizing stockholder welfare [and the] greater the extent of management stockholding, the more likely it is that the firm will adopt strategies consistent with maximizing stockholder wealth" (p. 579).

Hill and Snell's study confirms that "[stock] concentration had a strong impact upon strategy." There is a positive correlation between stock concentration and research and development expense, which suggests that "stockholders favored an emphasis upon innovation." There was a negative correlation between stock concentration and diversification, "suggesting that when stockholders were weak, managerial preferences for diversification dominated" (pp. 586-587).

Organizational Structure and Risk in R&D

Hokisson and Hitt (1988) accept that previous research establishes that the strategic control systems and incentives at M-form and U-form management levels result in different managerial risk preferences. Their "central argument is that adoption of multidivisional (M-form) structures and strategic control systems in highly diversified firms results in short-term-oriented risk preferences between both levels of managers" (p. 605). "M-form" is an abbreviation for multi-division form, describing an organization comprising atomistic work units grouped under a larger rubric in that each unit's work complements the others. This differs from "U-form," standing for unitary form, indicating an organization comprising specialized units conducting similar kinds of tasks and being grouped together.

The authors present three hypotheses: First is that "less diversified firms with U-form structures will invest relatively more in R&D than M-form firms. Further, highly diversified firms have lower strategic control of their various businesses and are not as likely to seek synergies among these businesses." Second, considering the inverse relationship between managers' "operational knowledge" of each unit, and the number of units within a diversified organization, "managers invest relatively less in R&D than in firms that are not as diversified." Third, the greater a firm's diversification, the more that firm will "use strategic control systems that are better designed to pursue hedging than synergy." In sum, Hokisson and Hitt suggest "that investment in R&D by highly diversified firms will not be evaluated as positively by the external capital market as it will be for less diversified firms" (p. 606). M-form firms can be characterized as maintaining acute financial control, suggesting "a short-term, low-risk orientation and thereby lower relative investment in R&D." The authors' study "of 124 major U.S. firms suggest[s] that less diversified U-form firms invest more heavily in R&D than more diversified M-form firms after controlling for size and industry effects." Moreover, the greater a firm's presence in a particular market, the more it will invest in research and development activity in comparison to "either related or non-related business firms" (p. 605).

Large M-form firms with diversified product lines are noted for their decentralized "operating responsibilities," which promote "the creation of two general managerial levels." The top level involves a general corporate office where top-level officers focus on overall strategic direction and resource allocation. The second level involves the top managers in separate business units (divisions) where the focus is on operational issues. This is in contradistinction to the tradition "centralized structures based upon functional departments (U- form systems) [in] M-form structures" (p. 606). The authors cite Williamson (1975), who suggests that "the M-form provides a number of beneficial features for managing diversity. These include:

1. Identification of distinct businesses within the firm;
2. Establishment of a division for each distinct business;
3. Decentralization of responsibility for operating each business to the appropriate division;
4. Centralization of overall strategic control at the general corporate office; and
5. Centralization of overall financial control and resource allocation at the head office" (Hokisson & Hill, 1988, p. 606).

Because large, diversified firms are increasingly dependent upon standardized performance criteria, "the M-form has become the dominant organizational form among large enterprises and ROI increasingly has become the major criterion for evaluating divisional contribution to company performance." The authors find that M-form management suffers from a myopic view on "measurable short-term efficiency." In an earlier study, Hokisson and Hitt found that even though the rate of return, over the long term, was not positively correlated to the adoption of M-form structures and behaviors, the "variability of rate of return decreased," suggesting that managers of M-form firms take fewer risks (p. 607). Hokisson and Hitt challenge the accepted position that Williamson presented in 1975, "that M-form controls limit opportunistic action by divisional managers." In fact, "M-form controls (beyond a certain span of control) may both foster divisional self-interest (through limitations on long-term investments such as R&D to bolster short-run performance) and allow its consequences to go undetected" (p. 608).

Hokisson and Hitt also discuss vertical and horizontal integration in relation to M-form structures and processes. Considering how leading firms use vertical integration to increase their efficiency and attain market share through horizontal acquisitions, "divisions often correspond to a stage of the production process requiring corporate coordination to manage product transfer effectively between upstream and downstream divisions" (p. 609). Furthermore, as horizontally integrated firms concentrate on specific competencies in specific markets or in related industries, suggesting a top-down structure, the notion that resources are allocated to distinct units based on their economic contribution to the firm becomes problematic; in fact, "The imposition of such centralized operational controls violates the principle of autonomy in the M-form" (p. 609).

Corporate Strategy vs. R&D Expenditure

Baysinger and Hoskisson (1989) provide "empirical evidence that choice of diversification strategy systematically affects R&D intensity in large multi-product firms," suggesting that R&D intensity is highest in "dominant-business" and "related-business firms" in comparison to unrelated-business firms. They determined that there is need to investigate "the implications of different types of corporate diversification strategy on the management of corporate-strategic-business-unit relationship [in] large multi-product corporations" (p. 310). The authors examined 971 firms that employed four different major diversification strategies: technological opportunity, current liquidity, debt position, and market structure, suggesting in accord with previous research that "dominant-business and related-constrained diversification strategies are most likely to be implemented with internal controls that foster risk taking and, hence, R&D spending, but that related-linked and unrelated diversification strategies are most likely to be implemented using internal controls that inhibit risk taking and, hence, R&D spending." Although they focused on "the empirical relation between corporate strategies and actual R&D expenditures," they believe that "a number of other economic factors may also influence those decisions" (pp. 311-312, 316-317).

After the Second World War, corporate "diversification...emerged as a major trend in the...U.S. economy." Building on research "addressing the question of why firms diversify" and the findings of Williamson (1975, 1985) and Chandler's (1962) original work, which "argued that firms virtually always implement diversification through the adoption of multi-divisional [M-form] structure and controls," the authors concur that "those characteristics should enhance firm performance, especially with respect to managers' willingness to assume risk and, hence, emphasize R&D." Modifying these findings slightly, Baysinger and Hoskisson (1989) favor "developed arguments suggesting that division managers operating within an M-form system of internal controls avoid risky strategies and instead sacrifice long-term investments in research and development to more immediate financial performance goals" (pp. 311-312). "Once an M-form structure is in place, different corporate diversification strategies may affect managerial risk propensity, as measured by R&D investment" (p. 313).

Previous research suggests that an "important criticism of diversified M-form firms is that division managers operating within their internal control systems develop short time horizons and thus avoid risky R&D investments, even when those investments promise a positive net expected value...In large diversified firms, corporate managers tend to use a return-on-investment (ROI) criterion for evaluating division managers' performance, causing division managers to meet short-term ROI objectives by reducing expenditures that are not essential for the attainment of short-run returns but are critical to the maximization of organizational efficiency in the long run." Moreover, it is believed that "M-form structure adversely affects top management's willingness to stress R&D as a basic component of competitive strategy...Rather than being simply a device for implementing corporate diversification strategy, [M-form structure] may also be a consolidation and control mechanism that is inherently conservative." In sum, "Large diversified firms thus may tend to provide environments hostile to innovation: the M-form structure may be inherently variance-reducing" (p. 314).

While it is accepted that "decomposition of strategic and operational decision making" and "decentralization of decision making so that it is carried out in semi-autonomous divisions...economize on the limited information-processing capacity of corporate managers...in a firm that has become highly diversified, both in terms of the number of divisions under top-management control and in terms of the lack of relatedness among product divisions, it seems highly unlikely that any given corporate manager will be adequately familiar with even a small percentage of the businesses in the firm's portfolio. Even for firms engaged in related diversification, top-level managers' ability to gather, process, and interpret the information needed to evaluate divisional performance accurately and allocate resources and rewards may be highly limited" (pp. 315-316). This does not mitigate the contrasting situation, in which "dominant-business firms and even in related-diversified firms that have engaged in only limited diversification" have a greater ability "to use both strategic and financial controls...because corporate managers cannot pay the same amount of attention to a large number of diverse divisions that they can to a smaller number of more related divisions" (p. 316). Ultimately, "firms implementing related-linked and unrelated strategies may maintain their efficiency in terms of production and information costs but may induce short-term, risk-averse behavior at the division level [of] the process" (p. 329).

Transaction Costs and the Nature of R&D Projects

In his 1990 *Administrative Science Quarterly* article, Pisano discusses two different sources of transaction costs—small-numbers-bargaining hazards and appropriability concerns—and how, if at all, they "affect established firms' choices between in-house and external sources of R&D when technological change shifts the locus of R&D expertise from established enterprises to new entrants, and established firms face a make-or-buy decision for R&D projects." Data analyzed were drawn from "92 biotechnology R&D projects that major pharmaceutical companies have sponsored either in-house or through external contractual arrangements." He presents two major hypotheses: "that small-numbers-bargaining hazards in R&D markets motivate internalization of R&D, [which] was supported by the data [and] that rivalry among established firms would lead to internalization, [which] was not supported" (Pisano, 1990, pp. 153, 174).

Previous research indicates that "there are many potential sources of transaction costs, including engineering intensity and design specialization (Monteverde & Teece, 1982a, 1982b; Masten, 1984), technological uncertainty (Walker & Weber, 1984; Balakrishnan & Wernerfelt, 1986), and the co-location of specialized assets (Joskow, 1987)." Pisano (1990) "focuses on the following two: (1) small-numbers-bargaining hazards stemming from specialized R&D capabilities and (2) appropriability problems arising from competition in product markets" (p. 154). Recognizing the necessary link between a firm's arsenal of knowledge and its ability to compete, and that "in-house R&D has traditionally been an important source of technical know-how for firms," Pisano suggests that "firms can tap the R&D capabilities of competitors, suppliers, and other organizations through such contractual arrangements as licenses, R&D agreements, and joint ventures." Additionally, given that "small-numbers-bargaining hazards play a central role in the transaction-cost theory of the firm," Pisano points out that "theoretical and empirical analysis of their effects has been largely confined to intermediate product markets." Pisano's empirical analysis of bio-tech R&D projects investigates "whether small-numbers-bargaining hazards also influence R&D markets," an issue that has received scant attention (p. 153).

As Pisano defines it, "transaction-cost theory attempts to explain why institutional structures other than markets are necessary for the efficient governance of economic activity. It assumes that, due to economies of specialization and the administrative and incentive limits of hierarchies, markets are a more efficient governance structure, unless a transaction is surrounded by special circumstances."

As with any entity or system challenged by uncertainty, investment is made in what we know. "Transaction-cost theory posits that...uncertainty over the terms of trade arises when the contingencies affecting the execution of the agreement are complex and difficult for the trading partners to understand, predict, or articulate." Once the market is clear of contingencies, partners may wish to renegotiate contractual terms, but renegotiation "represents a hazardous proposition for a party that has limited exchange alternatives. This situation, known as the small-numbers-bargaining problem, can occur when a firm invests in assets that are costly to transfer to alternative transactions or uses. Because such transaction-specific assets limit the firm's ability to switch partners, they make it vulnerable to opportunistic re-contracting" (p. 153).

Pisano suggests employing transaction-cost theory, as it "can help us to understand R&D boundary choices that occur in the wake of technological changes that make existing R&D capabilities obsolete but preserve capabilities needed to commercialize the new technology [and] vertically integrate despite the benefits of trading with external parties" (p. 157). If, as has been accepted, "the boundaries of the firm are determined by the trade-off between the transaction costs of using the market and the organizational costs of using hierarchies," firms "will adjust their boundaries when the trade-off between transaction costs and internal organizational costs change" (p. 160).

R&D AND ORGANIZATIONAL KNOWLEDGE

Knowledge Transfer in R&D

Demonstrating the extent to which technology and innovation management is interdisciplinary in terms of methodologies invoked and subject matter affected, Martin and Irvine's 1983 *Research Policy* article "present[s] a framework for assessing the relative contributions to scientific knowledge made by different research groups in the same discipline" (p. 62), with findings extendable to organizations whose primary function is research and knowledge transfer. The paper "establishes a methodology that, while not able to compare directly research centers in different specialties, is able to identify those centers that are amongst the international leaders in their own specialties, and which, if other factors are equal or indeterminate, should be given a relatively high priority in their claims for research funds" (p. 89). Their approach to assessing the impact and contribution of research on a specific field, one that focuses on variables attributable to research centers rather than individuals, derives from a much more generalized and comprehensive combination of practical factors concerning rising costs and decreasing budgets of research and development firms in all industries and disciplines. Their methodology is an attempt to answer how organizations should go about establishing "priorities between research groups competing for scarce funds," by identifying and evaluating "one of the most important pieces of information needed by science policymakers: ...an assessment of those groups' recent scientific performance" (p. 61).

With their focus on research centers rather than individual scientists, as the majority of funding goes to centers rather than specific projects or people, the authors' goals include identifying "a number of 'partial indicators'—that is, variables determined partly by the magnitude of the particular contributions, and partly by 'other factors'" and converge those factors: if "the partial indicators are to yield reliable results, then the influence of these 'other factors' must be minimized" (p. 61). Their methodology, the authors suggest, "overcomes many of the problems encountered in previous work on scientific assessment by incorporating the following elements:

1. The indicators are applied to research groups rather than individual scientists;
2. The indicators based on citations are seen as reflecting the impact, rather than the quality or importance, of the research work;
3. A range of indicators are employed, each of which focuses on different aspects of a group's performance;
4. The indicators are applied to matched groups, comparing 'like' with 'like' as far as possible;
5. Because of the imperfect or partial nature of the indicators, only in those cases where they yield convergent results can it be assumed that the influence of the 'other factors' has been kept relatively small (i.e., the matching of the groups has been largely successful), and that the indicators therefore provide a reasonably reliable estimate of the contribution to scientific progress made by different research groups" (p. 61).

Their findings, based on the "study of four radio astronomy observatories," and employing their "method of converging partial indicators...(publications per researcher, citations per paper, numbers of highly cited papers, and peer evaluation) are found to give fairly consistent results." The authors conclude that their results support the finding that not only can basic research be assessed, but also that differences among research centers in a specific discipline can be ascertained, urging that "the method of converging partial indicators can yield information useful to science policymakers" (p. 61).

R&D and Market Integration

In what may be viewed as being diametrically opposed to Martin and Irvine's starting point, Gupta et al. (1986) attempt to "bridge the literature gap by developing a conceptual framework for the study of R&D-marketing integration in the innovation process...based on a synthesis of the literature from marketing, organizational behavior, business strategy, research management, innovation, and new product management." The relationship between research and development and marketing integration is fundamental to a firm's business strategy and its approach to environmental uncertainty. "Factors related to organizational design and senior management support, along with the socio-cultural differences between R&D and marketing managers, can influence the level of integration achieved by an organization [and the level of] innovation success" (p. 7).

Specifically, the authors take into account "the views of Child (1972) on strategic choice and that of Lawrence and Lorsch (1969) on environmental complexity," as well as "Hage (1980, p. 423), [who] suggests, 'Sometimes there is a great deal of strategic choice and at other times a great deal of environmental constraint'" (Gupta et al., 1986, p. 9). In establishing their hypothesis "that if a firm's innovation strategy involves being 'first in' with new products, markets, and technologies, it is likely to require a greater degree of R&D-marketing integration," the authors add Freeman (1974, p. 255) and Parker (1978, p. 98) to their base, appropriating "six broad types of innovation strategies available to firms: offensive, defensive, imitative, dependent, traditional, and opportunist. These strategies are juxtaposed to the typology presented by Miles and Snow (1978), who "categorize organizational strategy as Prospectors, Analyzers, Defenders, and Reactors." Ultimately, "what differentiates one strategy from another are the firm's goals and the degree of familiarity with its new products, markets, and technologies (Cooper, 1983b; Crawford, 1980)" (Gupta et al., 1986, p. 9). Gupta, Rai, and Wilemon also cite Mintzberg (1979, p. 221) regarding the types of contingency factors giving rise to differences in organizational structure. They include "organizational age, size, and ownership, [but] the most important contingency factors are environmentally related, such as environmental stability, complexity, diversity, and hostility." Earlier studies by "Burns and Stalker (1961), Woodward (1965), Lawrence and Lorsch (1969), Khandwalla (1974), and others suggest that the greater the environmental uncertainty, the greater the specialization or differentiation within the organization" (Gupta et al., 1986, p. 10).

Environmental uncertainty, in fact, is the factor that most heavily influences the integration of research and development with marketing. Whether it is Galbraith and Nathanson (1978) presenting contingency theory as integral to organizational design, or Downey, Hellriegial, and Slocum (1975) pointing to a firm's ability to come up with varying levels of information processing requirements as a result of environmental uncertainty, this contextual gap in knowledge affects a firm's "ability to anticipate changes in competitors' strategies, consumers' new product requirements, technology, emergence of new competitive forces in the market, and new regulatory constraints on product performance and design" (Gupta et al., 1986, p. 9).

How an organization is structured affects how, during the innovation process, information is gathered, disseminated, and used. "The organizational structure, then, is a critical variable in determining the information processing potential between its various subunits and with the environment" (Gupta et al., 1986, p. 10). Specifically, Deshpande (1982) and John and Martin (1984) have "found a negative relationship between centralization and use of market research information, credibility, and utilization of plan output. Moreover, "Hage and Aiken (1967), Palumbo (1969), Blau (1973), Daft and Becker (1978), and Hage and Dewar (1973) have found negative correlations between centralization and innovative output" (Gupta et al., 1986, p. 11). In a more recent article, however, Hage (1980, pp. 186-187) "points out that centralization can have a positive effect on innovation. He explains that in the presence of change-values, centralization may be positively related to innovation," noting that "mechanical organizations, which have low rates of change, are also the places where radical innovation can occur, because they are more likely to have a crisis as well as a structure that is more tolerant of dictatorial practices" (Gupta et al., 1986, p. 11).

Generally, as Galbraith and Nathanson (1978) suggest, agility is critical: "there is not a 'best' organizational structure; rather, structure should be adapted to the requirements of the task and the task environment." Mintzberg (1979), following Alvin Toffler (1971), "suggests that this could be called 'ad hocratic'. This structure is similar to that found in Lawrence and Lorsch's (1969) plastic companies, Burns and Stalker's (1961) electronic firms, Woodward's (1965) unit and process producers, and in NASA as described by Galbraith (1973). Such structures are characterized by (1) little formalization, (2) selective decentralization, (3) mutual adjustment as a coordinating mechanism, and (4) decision-making power distributed among managers and non-managers" (Gupta et al., 1986, p. 11).

In presenting the new product development process from the R&D-marketing perspective, the authors synthesize theory and empirical studies in fields such as "marketing, organizational behavior, new product development, and research management" to suggest that "a firm's strategy and its perceived environmental uncertainty will influence the extent of R&D-marketing integration the company will ideally require," and "that certain factors related to organizational design and senior management support, along with socio-cultural differences in the orientations of R&D and marketing managers, will affect the level of integration that can be achieved by an organization" (p. 14).

Absorptive Capacity and Innovation

Cohen and Levinthal (1990) focus on how individuals and the firms that employ them identify, assimilate, and use information to fuel innovation. "Absorptive capacity" is the authors' term for this process. The level of absorptive capacity is derivative of "the firm's level of prior related knowledge" and the extent to which a firm is composed of individuals with diverse expertise. Cohen and Levinthal look at the nature and processes of the development of absorptive capacity in three main ways. First, "the factors that influence absorptive capacity at the organizational level"; second, "how an organization's absorptive capacity differs from that of its individual members"; and third, "the role of diversity of expertise within an organization." They believe that there is a cause-and-effect relationship between a firm's ability to develop absorptive capacity and the extent to which that firm will invest in specific areas of expertise. Cohen and Levinthal consider the connection between absorptive capacity and innovation to be "history- or path-dependent and argue how lack of investment in an area of expertise early on may foreclose the future development of a technical capability in that area." After a review of the literature, the authors offer "a model of firm investment in research and development (R&D), in which R&D contributes to a firm's absorptive capacity, and tests predictions relating a firm's investment in R&D to the knowledge underlying technical change within an industry." In sum, prior knowledge is the fuel that drives a firm's ability to identify new information, assign it value, and apply it to its commercial endeavors (p. 128).

Prior research suggests that the more a firm conducts its own research and development, the greater the use of external information it will achieve. To Cohen and Levinthal this means that absorptive capacity emanates from investment in R&D. Additionally, there is another view suggesting that "through direct involvement in manufacturing, a firm is better able to recognize and exploit new information relevant to a particular product market" (p. 129). The greater the access to information and new experiences in a particular domain—focused training sessions and intensive workshops, for example—the greater an individual's absorptive capacity can become. Creativity and absorptive capacity, from a cognitive psychological perspective, are similar. There are positive correlations between the depth of information processed, the amount of effort used in that processing, and the facility to make relevant associations between new information and what has already been internalized. The authors invoke Harlow's 1949 propositions concerning "learning-set theory, [suggesting] that important aspects of learning how to solve problems are built up over many practice trials on related problems" (Cohen & Levinthal, 1990, p. 131).

The extent to which any organization is absorptive is dependent on the "absorptive capacities of its individual members." Should an organization desire to increase that capacity, it would do well to inculcate practices that not only generate new information, but also "build on prior investment in the development of its constituent, individual absorptive capacities" because "organizational absorptive capacity will tend to develop cumulatively." Moreover, the information absorbed, to be valuable, must be put into use; therefore, "an organization's absorptive capacity does not simply depend on the organization's direct interface with the external environment. It also depends on transfers of knowledge across and within subunits that may be quite removed from the original point of entry," putting a spotlight on "the structure of communication between the external environment and the organization, as well as among the subunits of the organization, and also on the character and distribution of expertise within the organization." As the rate of change and information uncertainty increases, Cohen and Levinthal suggest that a centralized interface—a "gatekeeper" of information—will be an ineffective tool for increasing the capacity for information absorption. In fact, "it is best for the organization to expose a fairly broad range of prospective 'receptors' to the environment." When there is a group of people who have diverse knowledge bases and can communicate that knowledge effectively, there is opportunity for "effective communication both within and across subunits" (pp. 131-132).

One caveat presented is the case of "all actors in the organization shar[ing] the same specialized language [and therefore being] effective in communicating with one another, but [unable] to tap into diverse external knowledge sources." However, Cohen and Levinthal remind us that "diverse knowledge structures coexisting in the same mind elicit the sort of learning and problem solving that yields innovation." If we consider the same mind to be the collective knowledge of an organizational unit, communication becomes essential to innovation. The exchange of knowledge via effective communication among individuals "will augment the organization's capacity for making novel linkages and associations—innovating beyond what any one individual can achieve." In fact, "an organization's absorptive capacity is not resident in any single individual but depends on the links across a mosaic of individual capabilities" (p. 133).

Absorptive capacity, as has been noted, is more than the accretion of individual knowledge bases and effective inter- and intra-environmental communication. "To the extent that an organization develops a broad and active network of internal and external relationships, individuals' awareness of others' capabilities and knowledge will be strengthened. As a result, individual absorptive capacities are leveraged all the more, and the organization's absorptive capacity is strengthened" (p. 134). Absorptive capacity is also an assistive element in a firm's ability to seize "emerging technological opportunities...Thus, organizations with higher levels of absorptive capacity will tend to be more proactive, exploiting opportunities present in the environment, independent of current performance. Alternatively, organizations that have a modest absorptive capacity will tend to be reactive, searching for new alternatives in response to failure on some performance criterion that is not defined in terms of technical change per se (e.g., profitability, market share, etc.)" (p. 137).

The Mechanics of Knowledge Transfer

Lane and Lubatkin (1995) modify Cohen and Levinthal's construct, absorptive capacity, by focusing on three key elements of knowledge transfer: the type of knowledge offered by the "teaching" firm, the nature and amount of similarity between "teacher" and "student" firms in terms of compensation practices and organizational structures, and the level of familiarity the student firm has with the teacher firm's organizational problems. The authors construct a research model based on these elements and apply it to data gathered from "R&D alliances between pharmaceutical and biotechnology companies, using measures of a firm's scientific knowledge base and research capabilities, and measures of how similar the 'student' and 'teacher' firms are in those areas. Our measures of firm knowledge utilize a type of bibliometric data not widely used in strategy research" (Lane & Lubatkin, 1995, p. 462).

As cognitive science and learning researchers have found, individuals and businesses rely on prior knowledge of related entities to develop new information, new products, and new strategic understandings in competitive environments. Lane and Lubatkin look at intra-organizational learning, particularly the characteristics of learning partners, to further their understanding of this dynamic. Their approach is different from previous research on inter-organizational learning in that prior research concentrates on the role of absorptive capacity, which Cohen and Levinthal have defined as a firm's ability to value, assimilate, and utilize new external knowledge. Rather than thinking that learning can occur at similar levels between any given firm and all its domain competitors, the authors "reconceptualize the firm-level construct absorptive capacity as a learning dyad-level construct, relative absorptive capacity. One firm's ability to learn from another firm is argued to depend on the similarity of both firms' (1) knowledge bases, (2) organizational structures and compensation policies, and (3) dominant logics." Their findings included "similarity of the partners' basic knowledge, lower management formalization, research centralization, compensation practices, and research communities [that] were positively related to inter-organizational learning. The relative absorptive capacity measures are also shown to have greater explanatory power than the established measure of absorptive capacity, R&D spending" (p. 461).

Lane and Lubatkin stress the importance of learning alliances, dyads of firms, one termed the "teacher" and the other, the "student." Prior research has indicated that such alliances can accelerate the learning firm's ability to develop new knowledge and new products while minimizing "their exposure to technological uncertainties by acquiring and exploiting knowledge developed by others" (p. 461). The primary benefit of learning alliances is the enrichment of a firm's absorptive capacity. The authors refine Cohen and Levinthal's approach to inter-organizational learning by calling attention to the fact that "a firm's knowledge includes both easily communicated articulable knowledge and tacit knowledge, which is difficult to define due to its interconnections with other aspects of the firm such as its processes and social context" (p. 462).

New external knowledge can be acquired passively, actively, and interactively, with each method providing a different type of knowledge. Passive learning is analogous to traditional pedagogical methods, where students absorb knowledge from the texts they read, their instructors' lectures, and the tutorials they attend. Active learning, "such as benchmarking and competitor intelligence can provide a broader view of other firms' capabilities." The student partner in the learning dyad can acquire "the 'who, what, when, and where'." Liken this methodology to following instructions in a manual; the knowledge acquired does "not permit a firm to add unique value to its own capabilities." For this to happen, the dyad must engage in "interactive learning [in which] a student firm gets close enough to the teacher firm to understand not just the objective and observable components of the teacher's capabilities, but also the more tacit components: the 'how and why' knowledge" (pp. 462-463).

Lane and Lubatkin exploit prior research on cognitive structures and problem solving to support what some would accept as a truism, "that an individual's learning is greatest when the new knowledge to be assimilated is related to the individual's existing knowledge structure." This is the operative concept informing analogy, and it is closely related to Cohen and Levinthal's "absorptive capacity." The authors extend this construct to the concept of inter-organizational learning. Understanding fundamental concepts of a traditional discipline prepares for advances in both related areas and its sub-disciplines. The authors use the example of "a chemistry scholar [who] may not be able to appreciate advances in biotechnology without first having an understanding of basic biological sciences." Once a student firm recognizes external knowledge that is potentially valuable to its organization, it must take the next step and internalize it. Citing Cohen and Levinthal, Lane and Lubatkin (1995) point out that "the assimilation process is influenced by a firm's tacit, firm-specific knowledge regarding its established systems for processing knowledge" (p. 464). The authors use the example of rules-based computer programs and the operating systems they run on: software will not function on operating systems that do not recognize the operations of the software itself.

Lane and Lubatkin also discuss compensation practices as influencing a firm's ability to be innovative and solve problems "at both the divisional level and business unit level." The inference drawn by the authors is that "the similarity of two firms' compensation policies serves as one proxy for the similarity of their knowledge-processing systems and norms [which] suggests that the second dimension of absorptive capacity, the ability to assimilate new external knowledge, is in part a function of the relative similarity of the student and teacher firms' compensation practices." In addition to compensation practices, a firm's organizational structure—"the degree of formalization and centralization used by the firm when allocating tasks, responsibilities, authority, and decisions...is important to how firms process knowledge because organization members interact not only as individuals, but also as actors performing organizational roles." Understanding a firm's structure is important because structure indicates an organization's "perception of the environment...influences an organization's communication processes [and] is strongly related to an organization's problem-solving behaviors," including its ability to commercially apply it to achieve organizational objectives (p. 465).

Bringing Academic Research to Industrial Fruition

Mansfield, in his 1991 *Research Policy* article, breaks new ground in an attempt to "estimate the extent to which technological innovations in various industries have been based on recent academic research, and the time lags between the investment in recent academic research projects and the industrial utilization of their findings" (p. 1). Mansfield collected "data concerning the percentage of new products and processes that, according to the innovating firms, could not have been developed (without substantial delay) in the absence of recent academic research," and posits that there is a relationship between new product development and the percentage of revenue returned to research and development. Because R&D-intensive firms tend to stay closely informed by academic research, they also tend to execute more innovations derivative of that research than less R&D-intensive firms. One of the benefits of staying abreast of academic research is economic: the more informed, the less expensive and time consuming (pp. 3, 11).

Maintaining Business Prominence in Technology Innovation

Christensen and Bower (1996) offer a view into technological innovation, realized through research and development, that seeks to address a fundamental problem: why established firms providing technological goods to an ever-growing marketplace falter in terms of new product development. The authors base their findings in an extensive examination of the international disk drive industry, and present a model "that charts the process through which the demands of a firm's customers shape the allocation of resources in technological innovation—a model that links theories of resource dependency to resource allocation." Industry leadership in technological development derives from an organization's understanding of existing customer needs. The questions arise: "Why and under what circumstances [do] financially strong, customer-sensitive, technologically deep and rationally managed organizations...fail to adopt critical new technologies or enter important markets—failures to innovate which have led to the decline of once-great firms?" (p. 197).

Christensen and Bower determine that "a primary reason why such firms lose their positions of industry leadership when faced with certain types of technological change has little to do with technology itself—with its degree of newness or difficulty, relative to the skills and experience of the firm. Rather, they fail because they listen too carefully to their customers—and customers place stringent limits on the strategies firms can and cannot pursue" (p. 198). While the authors conclude that "when significant customers demand it, sufficient impetus may develop so that large, bureaucratic firms can embark upon and successfully execute technologically difficult innovations—even those that require very different competencies than they initially possessed," the emphasis is on "may." Firms intent on innovation in rapidly changing and highly competitive technological markets also need to pay strict attention to proven, existing customer need, not just needs and desires they anticipated (p. 199).

Christensen and Bower define "technology" as "the processes by which an organization transforms labor, capital, materials, and information into products or services." Given that "most proposals to innovate require human and financial resources," companies often "mirror to a considerable degree the patterns in how its resources are allocated to, and withheld from, competing proposals to innovate" (p. 198). Quite often, existing leaders of innovative technologies "failed to develop simpler technologies that initially were only useful in emerging markets, because impetus coalesces behind, and resources are allocated to, programs targeting powerful customers. Projects targeted at technologies for which no customers yet exist languish for lack of impetus and resources." Firms that bring new emerging technologies to established markets more rapidly than those product demands can be assessed often challenge existing marketplace leaders for market share by staying invested in the new product as the emerging market becomes established. The authors ascribe leaders' failure to "managerial myopia or organizational lethargy, or to insufficient resources or expertise" (p. 197).

Firms that are resource dependent look "outside the firm for explanations of the patterns through which firms allocate resources to innovative activities" (p. 198). An alternative view suggests that "most strategic proposals—to add capacity or develop new products or processes—take their fundamental shape at lower levels of hierarchical organizations." The authors conjoin these ideas to demonstrate that "whether sufficient impetus coalesces behind a proposed innovation is largely determined by the presence or absence of current customers who can capably articulate a need for the innovation in question." Indeed, the authors demonstrate the causal connections stemming from customer expectation for product improvement, to the nature of new proposals and their relationship to a firm's technological capabilities, to the types of marketplaces targeted to infuse innovative technologies, and ultimately to a firm's "ultimate commercial success or failure with the new technology" (p. 199).

While Christensen and Bower are in line with those who agree with resource dependence theorists, those "who contend that a firm's scope for strategic change is strongly bounded by the interests of external entities...who provide the resources the firm needs to survive," their findings "decidedly do not support a contention that managers are powerless to change the strategies of their companies in directions that are inconsistent with the needs of their customers as resource providers...managers can, in fact, change strategy—but...they can successfully do so only if their actions are consistent with, rather than in counteraction to, the principle of resource dependence" (p. 212). Ultimately, "while many scholars see the issue primarily as an issue of technological competence, we assert that at a deeper level it may be an issue of investment. We have observed that when competence was lacking, but impetus from customers to develop that competence was sufficiently strong, established firms successfully led their industries in developing the competencies required for sustaining technological change. Importantly, because sustaining technologies address the interests of established firms' existing customers, we saw that technological change could be achieved without strategy change" (p. 215).

CONCLUSION

The seminal articles on the management of R&D consider a variety of different issues that relate to the uncertainty inherent in R&D. The effect of firm financial structure (Guerard et al., 1987) and ownership structure (Hill & Snell, 1988) are found to be important to decisions regarding investment in R&D. Similarly, organizational structure has a strong influence on the nature of R&D (Hoskinsson & Hitt, 1988), and R&D intensity is higher in less diversified businesses (Baysinger & Hoskisson, 1989). This body of research offers such insights as that a centrally controlled firm with an owner-operator is more likely to conduct research leading to innovations that produce substantial profit. Alternately, a multi-divisional diversified firm with a professional management structure is likely to conduct low-risk, low-return research. The emerging issue of outsourcing R&D, or acquiring R&D from external sources, raises new management issues such as transaction costs (Pisano, 1990) as well as concerns over knowledge transfer and appropriability (Cohen & Levinthal, 1990). In other words, a firm that attempts to acquire knowledge from external sources needs a certain amount of internal knowledge to be able to grasp the value and complexity of the knowledge that is available. For knowledge transfer there are three key elements: (1) the type of knowledge offered by the "teaching" firm, (2) the similarity between "teacher" and "student" organizations with respect to compensation practices and organizational structures, and (3) the level of familiarity the student organization has with the teacher organization's internal structure (Lane & Lubatkin, 1995).

The need to balance the different perspectives of R&D as a manner of commercializing science as opposed to solving customer problems is nicely developed by work discussing the prominence of research centers (Martin & Irvine, 1983) vs. the customer R&D interface (Gupta et al., 1986).We learn that the research center should be the unit of analysis, not the individual researcher, and are offered techniques to assist in benchmarking and assessment (Martin & Irvine, 1983). While from a customer perspective, we see that R&D and marketing should be closely aligned, the problem of leaning too heavily on the customer-R&D interface is especially apparent in times of major technological change and/or disruption to the technological base to existing markets (Christensen & Bower, 1996). This problem is most often felt by established firms. We find that R&D-intensive firms that stay abreast of developments, through technology scanning and internal research activities, are best positioned to respond quickly and effectively to threats and opportunities that arise. Finally, the use and lack of use of various management science techniques by R&D managers (Liberatore & Titus, 1983) has been of great importance in assisting theoreticians in better identifying the needs of R&D managers and the opportunities for the application of new and existing theory to the management of R&D.

REFERENCES

Balakrishnan, S., & Wernerfelt, B. (1986). Technical change, competition, and vertical integration. *Strategic Management Journal, 7,* 347-359.

Baysinger, B., & Hoskisson, R.H. (1989). Diversification strategy and R&D intensity in multi-product firms. *Academy of Management Journal, 32*(2), 310-332.

Berle, A., & Means, G. (1932). *The modern corporation and private property.* New York: Commerce Clearing House.

Blau, P. (1973). *The organization of academic work.* New York: John Wiley & Sons.

Burns, T., & Stalker, G. (1961). *The management of innovation.* London: Tavistock.

Chandler, A. (1962). *Strategy and structure: Chapters in the history of American industrial enterprise.* Cambridge, MA: MIT Press.

Child, J. (1972). Organizational structure, environment, and performance: The role of strategic choice. *Sociology, 6*(January), 2-22.

Christensen, C.M., & Bower, J.L. (1996). Customer power, strategic investment, and the failure of leading firms. *Strategic Management Journal, 17*(3), 197-218.

Cohen, W.M., & Levinthal, D.A. (1990). Absorptive capacity—a new perspective on learning and innovation. *Administrative Science Quarterly, 35*(1), 128-15.

Cooper, R. (1983). The impact of new product strategies. *Industrial Marketing Management–EM-30, 1*(February), 2-11.

Crawford, C. (1980). Defining the charter for product innovation. *Sloan Management Review, 22*(Fall), 3-12.

Daft, R., & Becker, S. (1978). *Innovation in organizations: Adoption in school organizations.* New York: Elsevier.

Deshpande, R. (1982). The organizational context of market research use. *Journal of Marketing, 46*(Fall), 91-101.

Dhrymes, P., & Kurz, M. (1964). On the dividend policy of electric utilities. *Review of Economics and Statistics, 46,* 76-81.

Downey, G., Hellriegial, D., & Slocum, W. (1975), Environmental uncertainty: The construct and its application. *Administrative Science Quarterly, 20*(December), 618-629.

Freeman, C. (1974). *The economics of industrial innovation.* Baltimore, MD: Penguin.

Galbraith, J. (1973). *Designing complex organizations.* Reading, MA: Addison-Wesley.

Galbraith, J., & Nathanson, D. (1978). *Strategy implementation: The role of structure and process.* St. Paul, MN: West.

Galbraith, C., Samuelson B., Stiles C., & Merrill, G. (1986). Diversification, industry research and development, and market performance. *Academy of Management Proceedings,* 17-20.

Guerard, J.B., Bean, A.S., & Andrews, S. (1987). R&D management and corporate financial policy. *Management Science, 33*(11), 1419-1427.

Gupta, A.K., Rai, S.P., & Wilemon, D. (1986). A model for studying R&D—marketing interface in the product innovation process. *Journal of Marketing, 50*(2), 7-17.

Hage, J. (1980). *Theories of organizations.* New York: John Wiley & Sons.

Hage, J., & Aiken, M. (1967). Program change and organizational properties: A comparative analysis. *American Journal of Sociology, 72*(March), 503-519.

Hage, J., & Dewar, R. (1973). Elite values versus organizational structure in predicting innovation. *Administrative Science Quarterly, 18*(September), 279-290.

Harlow, H. (1949). The formation of learning sets. *Psychological Review, 56,* 51-65.

Hill, C.W.L., & Snell, S.A. (1988). External control, corporate strategy and firm performance research-intensive industries. *Strategic Management Journal, 9*(3), 577-590.

Hokisson, R.E., & Hitt, M.A. (1988). Strategic control-systems and relative R-and-D investment in large multiple product firms. *Strategic Management Journal, 9*(6), 605-621.

John, G., & Martin, J. (1984). Effects of organizational structure of market planning on credibility and utilization of plan output. *Journal of Marketing Research, 21*(May), 170-183.

Joskow, P. (1987). Contract duration and relationship-specific investments: Empirical evidence from coal markets. *American Economic Review, 77,* 168-185.

Khandwalla, P. (1974). Environment and its impact on the organization. *International Studies of Management Organization, 2*(Fall), 297-313.

Kay, N. (1984). *The emergent firm.* New York: St. Martin's Press.

Lane, P., & Lubatkin, M. (1998). Relative absorptive capacity and interorganizational learning. *Strategic Management Journal, 19*(5), 461-477.

Lawrence, P., & Lorsch, J. (1969). *Organization and environment: Managing differentiation and integration.* Homewood, IL: Irwin.

Liberatore, J., & Titus, G.J. (1983). The practice of management science in R&D project management. *Management Science, 29*(8), 962-974.

Mansfield, E. (1991). Academic research and industrial innovation. *Research Policy, 20*(2), 1-12.

Marris, R. (1964). *The economic theory of managerial capitalism.* London: Macmillan.

Martin, B.R., & Irvine, J. (1983). Assessing basic research: Some partial indicators of scientific progress in radio astronomy. *Research Policy, 12*(3), 61-90.

Masten, S. (1984). The organization of production: Evidence from the aero-space industry. *Journal of Law and Economics, 27*(October), 403-417.

McCabe, G. (1979). The empirical relationship between investment and financing: A new look. *Journal of Financial and Quantitative Analysis, 14,* 119-135.

Miles, R., & Snow, C. (1978). *Organizational strategy: Structure and process.* New York: McGraw-Hill.

Mintzberg, H. (1979). *The structure of organizations.* Englewood Cliffs, NJ: Prentice Hall.

Monteverde, K., & Teece, D. (1982a). Supplier switching costs and vertical integration. *Bell Journal of Economics, 13,* 206-213.

Monteverde, K., & Teece, D. (1982b). Appropriable rents and quasi-vertical integration. *Journal of Law and Economics, 25*(October), 321-328.

Palumbo, D. (1969). Power and role specificity in organizational theory. *Public Administrative Review, 29*(May-June), 237-248.

Parker, J. (1978). *The economics of innovation—the national and multinational enterprise in technological change.* New York: Longman.

Pisano, G.P. (1990). The R&D boundaries of the firm: An empirical analysis. *Administrative Science Quarterly, 35*(1), 153-176.

Toffler, A. (1971). *Future shock.* New York: Bantam.

Walker, G., & Weber, D. (1984). A transaction cost approach to make-or-buy decisions. *Administrative Science Quarterly, 29,* 373-379.

Williamson, O.E. (1985). *The economic institutions of capitalism: Firms, markets, and relational contracting.* New York: Macmillan Free Press.

Williamson, O.E. (1975). *Markets and hierarchies: Analysis and antitrust implications.* New York: Macmillan Free Press.

Woodward, J. (1965). *Industrial organization: Theory and practice.* London: Oxford University Press.

Chapter III
Technology Development and Innovative Practice

INTRODUCTION

This chapter on innovative practice supporting technological development has several thematic overlays that show some consistency in terms of patterns, but also some diversity in terms of strategies that researchers have employed in this area. Beginning with Hage and Aiken's (1969) seminal work on routinization and how the social structures of organizations affect technological development and innovation, readers will see two general trends in terms of approach: the statistical and the sociological. Whether it is Aldrich's (1972) use of path analysis to study the nature and effects of organizational variables on innovative practice, or Rothwell et al.'s (1974) identification of innovation success factors, or Downs and Mohr's (1976) defining of innovation through factors of variability, quantitative methods are shown to be increasingly powerful tools in identifying the nature of innovation and technology development. Nelson and Winter (1977) continue in this vein by establishing an inclusive theoretical structure for innovation, Dewar and Hage (1978) identify variables of structural differentiation and complexity that affect this domain, and Kimberly and Evanisko (1981) suggest variables to follow that come from both within individual organization units and their wider contexts. Pavitt (1984) uses sectoral pattern analysis to describe how a combination of technology sources, user requirements, and potential technology appropriation affect how we understand technical change and the structural relationships between technology and industry. Fisher and Fry (1971) end the quantitatively based section with a discussion of their substitution forecasting model.

From a more qualitative orientation, Abernathy and Clark (1985) introduce "transilience," or a set of categories of technological change that is aligned with evolutionary developments that are altered by varying managerial environments. Anderson and Tushman (1990) continue on the evolutionary track with an explanation of their cyclical model of technological change. Their model shares some basic affinities with Clark's (1985) evolutionary based view of technological change shaped by customer demands. Barley (1986, 1990) presents two sociologically oriented sets of ideas, one examining patterns of action and interaction, and the other presenting the benefits of examining the interaction between social action and social form in technologically innovative organizations. Dosi (1982), concluding this chapter, invokes the Kuhnian paradigm to demonstrate the strength of identifying patterns of continuous changes and moments of discontinuity in technologically innovative environments.

QUANTITATIVE PERSPECTIVES ON TECHNOLOGY DEVELOPMENT AND INNOVATIVE PRACTICE

Fundamental Voices in Organizational Analysis

Hage and Aiken (1969) are two of the earliest researchers to examine the role of technology as an explanatory element in organizational analysis. Previous researchers view technology as contributing to "different levels of alienation in American industry" and affecting "different aspects of the organization's structure and goals." Their paper explores the "connection between routine work [involving technology], organizational structure, and goals" as it appears in "people-processing organizations," as opposed to "continuous process or assembly-line" workflows (pp. 366-367). What are the relationships among "the degree of routineness of work, and the social structure and goals of health and welfare organizations"? Their analyses of data related to "sixteen social welfare and health organizations located in a Midwestern metropolis in 1967" indicate that the "social structure of organizations with more routine work are found to be more centralized, more formalized...but no relationship with stratification is found. Organizations with routine work are further found to emphasize goals of efficiency and the quantity of clients served, not innovativeness, morale, or quality of client services" (p. 366).

A routine workflow is defined as one in which "clients are stable and uniform and much is known about the particular process of treatment" (p. 366). The degree of routineness is a measurable dimension of technology "that can be applied equally to people-processing, industrial, and other kinds of organizations," as well as "provide the basis for general propositions that can be tested in many organizational contexts." This is so as a result of "the recognition that the work processes of an organization provide the foundation upon which social structure is built," and as such, "technology should influence the nature of that structure" (p. 367). Understanding the variety of job tasks, organizational configurations, and workflow patterns, Hage and Aiken used the organization as the unit of analysis, aggregating data from each "in order to calculate scores for measures of organizational structures" (p. 368). They sought to ascertain levels of "routineness: how much variety there is in work; job codification: how well defined the job is; rule observation: job specificity description; [and] enforcement of the rules: how concrete the job procedural manual is" (p. 368).

Hage and Aiken base their interpretations of the consequences of the degree of routineness of work on the nature of how technology affects social structure, and whether coordination of technology and task can "occur either via planning, programmed interaction, or feedback…If technology can be routinized, then coordination can be and probably will be planned and programmed. If it can't, then coordination must be effected via feedback" (p. 370). In the case of organizations in which the work situation places its members into situations that have varied client needs, "then greater organizational power will accrue to organizational members who interact with the clients most frequently." In other words, "the more routine the workflow, the greater the centralization of decision-making about basic organizational issues" (p. 370). Alternatively, when there is "both variability in clients as well as lack of knowledge about their handling, the power structure should be polycentralized" (p. 371).

Routine contributes positively to the process of formalization—the less variety in tasks and demands, the greater the manageability of tasks such as developing documentation in the form of "policy manuals, job descriptions or evaluation procedures" (p. 371). As the technology used to accomplish tasks becomes routinized, there is also "centralization of power, formalization of roles, and some lessening of the level of professionalization in the organization," which restricts the range of goals attempted and fulfilled. In such formalized and routinized organizations, efficiency and quantity trumps morale and quality. Concomitantly, "the routine organization's [concern] with stability and high profits—achieved via quantity of production and an avoidance of innovation"—is opposed by the non-routine organization's emphasis on "growth, quality, and innovation, being less concerned with making profits" (p. 373).

Hage and Aiken's analysis of their data results in the lack of support for "the argument that the organization that is emphasizing new programs is one interested in innovativeness." In fact, they found no statistically significant correlation to support such an assumption, as the data do not "demonstrate a correlation between the emphasis on the development of new programs and the routineness of work.

Admittedly, the development of new programs is not the only way in which organizations can innovate" (p. 374).

The Benefits of Path Analysis

Aldrich (1972) seeks to clarify what he finds to be obscured by other organizational theorists: whether technology should be considered an independent or dependent variable when studying the roles and effects of organizational variables. In other words, should technology (as it existed in 1972) be considered as something that a researcher or experimental scientist actively controls (independent), or should technology be something that changes as a result of other variables being manipulated? Once that question is addressed, he describes the usefulness of "path analysis for studying organizational variables" (pp. 26-27). By describing path analysis, "an investigator is forced to bring his assumptions out into the open" (p. 28).

Specifically, Aldrich takes another look at the Aston group findings, which show that technology is of minimal importance when compared to other organizational variables such as size, and he employs path analysis as his method to review their findings. This leads Aldrich to propose that technology should be treated as an independent variable while taking "a multivariate functional approach in both context and structure" in which contextual variables are treated as independent, and structural as dependent (p. 27). Using path analysis, investigators can measure the effect of "inter-correlated exogenous variables," or variables that are extrinsic to the subject of analysis, as a way to explain both direct and indirect effects. "Path analysis does not allow the investigator to assign all of the inter-correlation effect to one of the two or more variables, of course, but its virtue lies in showing that, in the absence of more knowledge about the exogenous variables, such a procedure would be illegitimate." Path analysis opens the door to examining whether or not technology "is not simply acting as a dummy variable for" manufacturing or non-manufacturing organizations (p. 28). Rather than dismiss technology as immaterial or of lesser significance to organizational theory as previous researchers had done, as they focus on size of organization, Aldrich explains that path analysis does not allow for reliance on the "logical implications of their implicit causal model" or permit "the failure of the investigators to search for alternative interpretations and arrangements of the causal relationships among the variables studied." Returning to the Aston group as a reference point, Aldrich points out that their "follow-up study of fourteen organizations over a four- to five-year period...found no association between changes in size and changes in the structuring of activities" (p. 40).

Innovation and Factors of Success and Failure

Rothwell et al. (1974) report on the second phase results of a comparative analysis of commercially successful and commercially unsuccessful paired technological innovations in the chemical processing and scientific instruments industries. Whereas Phase 1 of the SAPPHO project involved 29 pairs, Phase 2 expanded the study to 43 pairs, 22 in the chemical processes and 21 in scientific instruments. Their goal was to "assess the value of user needs in technology implementation," and Phase 2 confirms the findings arising from their analysis of Phase 1 data. The data also elucidate "inter-industry differences [relating] to basic structural and environmental differences which exist between the two industries." The investigators looked further into the 34 cases of failure to ascertain the factors "which contributed maximally to the individual failures," as well as offering hypotheses to explain innovative success (p. 258). Rothwell et al. (1974, pp. 259-260) identify five principal areas of difference among both industries that separate success from failure in innovation:

1. Successful innovators were seen to have a much better understanding of user needs.
2. Successful innovators pay more attention to marketing and publicity.
3. Successful innovators perform their development work more efficiently than failures, but not necessarily more quickly.
4. Successful innovators make more use of outside technology and scientific advice, not necessarily in general but in the specific area concerned.
5. The responsible individuals in the successful attempts are usually more senior and have greater authority than their counterparts who fail.

The authors distinguish, however, between the narrow definition of commercial success in innovation as simply growth in revenue and market share, and the more complex or comprehensive meaning, where "the overall success of an innovation must be measured by the total impact of the innovation on the innovating organization" (p. 269). Ultimately, need satisfaction is the key factor: "User needs must be precisely determined and met, and it is important that these needs are monitored throughout the course of the innovation since they very rarely remain completely static" (p. 289).

Innovation and Instability

Downs and Mohr's 1976 contribution to the literature on technology and innovation in organizations is offered not to construct a specific theory of innovation, but to clarify the pertinent and different senses of innovation as a term describing complex organizations. Innovation is most often defined as the adoption of means or ends that are new to an organization. Researchers can analyze innovation in terms of the relative innovativeness of an organization or from the perspective of a specific method or tool's adoptability. The authors suggest that these two perspectives are "closely related methodologically and that this relationship describes an important symmetry" (p. 701). Their research involves identifying the factors that create instability in empirical studies and analyzing those factors in the context-complex organizational settings. Downs and Mohr "define four primary sources of instability: (1) variation among primary attributes, (2) interaction, (3) ecological inferences, [and] (4) varying operationalization of innovation" as the bases for their conclusions, which are intended to support "the development of an integrative theory" of innovation adoptability and organizational innovativeness (p. 701).

Fundamentally, the term innovation carries with it positive value and connotations. The difficulty in arriving at a general theory of innovation resides in the fact that the means and ends associated with innovation are not universal—some organizations determine that a method for improvement or tool that has not been employed by them would be unsuitable, while others would view the same method or tool as potentially valuable in attaining organizational improvements. This instability stymies the development of theories; to remedy this situation, the authors suggest that we investigate the sources of instability themselves, "reject the notion that a unitary theory of innovation exists and postulate the existence of distinct types of innovations whose adoption can best be explained by a number of correspondingly distinct theories" (p. 701). A beginning step would be to create a typology of innovation by separating attributes into two camps: primary and secondary qualities. Classifications schema are not new, of course; they are at the heart of positivist science. Following Locke, when "a typology is based on a primary attribute, an innovation can be confidently classified without reference to a specified organization. Regardless of its size, wealth, complexity, decentralization, and so forth, each organization would place the innovation in the same cell of the typology…Secondary qualities are those which are perceived by the senses, and so may be differently estimated by different percipients; primary qualities are those which are essential to the object or substance and so are inherent in it whether they are perceived or not." Primary innovations would be those that all types of organizations would agree to be innovations, but "when a typology is based on a secondary attribute, the classification of the innovation depends on the organization that is contemplating its adoption" (p. 702). Keep in mind, however, that "variation of a characteristic of an innovation between one study and another but not within the respective studies is almost certainly an important source of instability in innovation research" (p. 703). Recognizing that "determinants of adoption are different for different categories of innovations...innovation may be classified in different categories for different organizations; for example, an innovation might be seen as minor or routine by some organizations but as major or radical by others" (pp. 703-704). While these secondary attribute typologies affect research design, they should not be construed as "a liability," but more an indication that "we must build the idea of statistical interaction into our models of innovation...If we are studying an innovation that would be a reorientation for some organizations and a variation for others, all we need do is insert a variable which measures how compatible the innovation is to each organization in the sample. This would provide us with even more information than we would obtain from separate studies of variations and reorientations" (pp. 704-705).

Downs and Mohr (1976, p. 705) suggest that implementing these typologies will help "obtain the following four types of information:

1. By including the organization's compatibility with the innovation as a separate independent variable, we can determine the extent to which this in itself affects the adoption of the innovation—note that to make compatibility an independent variable we must transform it from a property of the innovation to a property of the organization, and that this can easily be done.
2. By employing one or more interaction terms—for example, the product of compatibility and executive ideology—we can determine the differential importance of a variable, such as executive ideology, for innovation, given any particular level of compatibility.
3. We can also investigate which characteristic or set of characteristics of the organization determines the classification of the innovation.
4. If, for the sake of discussion, size is found to play such a determining role, then we might also investigate the interaction of size and executive ideology as predictors of innovation. This may be an interesting and prescriptively valuable finding that could not be expected to emerge from the separate study of variations and reorientations."

If we think of secondary attributes not as being composed wholly of characteristics of the innovation or the organization, but as characterizing the relationship between the two, "the unit of analysis is no longer the organization but the organization with respect to a particular innovation...no longer the innovation, but the innovation with respect to a particular organization." The value of this approach can be realized by employing "an innovation-decision design, a consideration of the unit of analysis as an organization in relation to an innovation." Rather than approaching an organization and its ability to adopt innovation with the assumption that it is an unyielding, static entity, this approach keeps our attention on "the shifting incentives and constraints that are relevant to the decision to innovate" (p. 706). The innovation-decision design approach, as opposed to the multivariate approach of "treating several innovations...as an aggregate...considers them as discrete units, thereby preserving the special theoretical implications of each" (p. 708).

Downs and Mohr (1976, pp. 713-714) propose that researchers:

1. Use studies of different innovations to expose the impact of primary-attribute variation on models of innovation. This will involve observing and reporting the primary attributes of innovations and restricting generalizations from a given study to innovations in the same category of a primary-attribute typology rather than expecting all results to be identical.
2. Measure the secondary attributes of innovations (compatibility, relative advantage, and so forth) with respect to each organization and consider them as characteristics of adopters.
3. Use interactive models.
4. Use the innovation-decision design as the basis for analysis.
5. Do not conduct multiple-innovation studies in which the organization is assigned an aggregate score for innovation.
6. Recognize that extent of adoption and time of adoption are distinct concentralizations of innovation. Do not generalize from one dependent variable to the other. Do not use either as a comprehensive measure of innovativeness. There is not a single, unitary theory, but rather different theories to explain different aspects of innovation.
7. Study the adoptability of innovations by using either many innovations in relation to one single organization or by using the innovation-decision design.

An Inclusive Theoretical Structure for Innovation

Nelson and Winter (1977) contribute to establishing "some directions that would seem fruitful to follow if we are to achieve a theoretical structure that can be helpful in guiding thinking about policy" when the subject of that policy is innovation, broadly defined by the authors as "the wide range of variegated processes by which man's technologies evolve over time" (p. 37). Their focus is on developing theory that will address "the vast inter-industry differences in rates of productivity growth, and other manifestations of differential rates of technological progress across industries," arguing that attention should be paid primarily to those sectors of commerce that are lagging in an effort to improve their status. Current theory, in their estimation, fails because it is "fragmented, and knowledge and research fall into a number of distinct intellectual tradeoffs." Moreover, current theory neglects the fact that "innovation involves uncertainty in an essential way, and that the institutional structure supporting innovation varies greatly from sector to sector." Therefore, the authors offer "a theoretical structure that appears to bridge a number of presently separate subfields of study of innovation, and which treats uncertainty and institutional diversity centrally" (p. 36).

Nelson and Winter begin by acknowledging the complexities of establishing the valences of causes for innovation uncovered in the relationships between research and development, productivity, pricing, output, scale economies—even labor relations: these are some of the factors contributing to the paucity of theory applicable to policy decisions. The authors identify two "possible explanations" for this disconnect. "One is that research and development activity is more powerful when directed toward the technologies of certain industries than toward the technologies of others; therefore, the disparities in rates of technical progress reflect some kind of innate differences on ability to advance efficiently the different kinds of technologies. The second possible explanation (not mutually exclusive) focuses not on possible innate differences in what research and development can do in different sectors, but on differences in institutional structure that influence the extent to which research and development spending is optimal and the results of research and development effectively employed" (p. 45).

Nelson and Winter suggest that there are wide differences among firms regarding how "the results of research and development spending are internalized...that in some industries but not in others there is significant government subsidization of research and development where externalities are important, and that industries also differ significantly in the speed and reliability of the mechanisms by which new technology is screened, and the use of efficacious innovation spread throughout the sector" (p. 45).

Recognizing that innovation and uncertainty are inextricable, "a theoretical structure must encompass an essential diversity and disequilibrium of choices" (p. 47). We therefore need to keep uncertainty in the forefront when constructing policy, and understand that a variety of innovations is desirable, but it is incumbent upon an organization to maintain awareness of which innovations present the possibility of positive ends, and shunt aside those initiatives that show signs of failure. More specifically, "the institutional structure for innovation often is quite complex within an economic sector, and varies significantly between economic sectors," calling for "sub-theories of the processes that lead up to a new technology ready for trial use, and of what we call file selection environment that takes the flow of innovations as given" (pp. 47, 49).

Previous research often focuses on one of two "classes of factors influencing the allocation of effort: factors that influence the demand for or pay-off from innovation, and factors that influence the difficulty or cost of innovation," with the demand side suggesting "a simple model in which changes in composition of demand for goods and services across industries chain back to influence investment patterns, which in turn influence the relative return to investors working on improvements in different kinds of machines" (p. 49). Research on the supply side of the equation is focused on the "differences in the difficulty or cost of different kinds of innovation [have] had but [offer] limited conceptual and empirical pay-off" (pp. 49-50). One difficulty with the "tendency of some authors to try to slice neatly between invention and adoption, with all of the uncertainty piled on the former, [is that] one cannot make sense of the micro studies of innovation unless one recognizes explicitly that many uncertainties cannot be resolved until an innovation actually has been tried in practice" (p. 61).

Nelson and Winter assert that "successful innovation leads to both higher profit for the innovator and to profitable investment opportunities. Thus profitable firms grow. In so doing they cut away the market for the innovators and reduce their profitability, which, in turn, will force these firms to contract. Both the visible profits of the innovators and the losses experienced by the laggards stimulate the latter to try to imitate" (p. 64). There is, however, a difference between product and process innovation. Consumer reaction to new products has a direct connection to profitability, whereas for process innovation, "The firm can make an assessment of profitability by considering the effects on costs, with far less concern for consumer reaction" (pp. 64-65). This dichotomy between product and process is mirrored in theories about "market selection environments," which separate producers from consumers and regulators. "Consumers' evaluation of products—versus each other, and versus their price—is presumed to be the criterion that ought to dictate resource allocation. Firms can be viewed as bidding, and competing, for consumer purchases, and markets can be judged as working well or poorly depending on the extent to which the profitability of a firm hinges on its ability to meet consumer demands as well as or better than its rivals" (p. 67).

Structural Difference and Complexity

In their 1978 paper, Dewar and Hage attempt to "synthesize much of the literature on technology and size relative to the two dependent variables that appear to be most alike: structural differentiation and complexity," by determining "whether size has the same impact on complexity that it appears to have on structural differentiation" and suggesting "hypotheses relating technology first to complexity and then to structural differentiation." Their findings include "that the most important determinant of differentiation in the division of labor is the scope of an organization's task, a technological dimension, and not organizational size" and "support only the inference of a moderate causal connection between both size or task scope and either form of differentiation" (pp. 111-112).

Structural differentiation and complexity are staples of the theoretical literature of technology innovation management, with studies consistently finding "that size is related to structural differentiation, but the relationship between size and complexity is less clear." When measuring structural differentiation, researchers look at organizational elements such as "job titles, number of departments, and number of levels," but when the subject is complexity, the elements under scrutiny include those akin to activity and learning, such as "the number of different occupations, level of training, and extent of professional activity" (p. 111). Dewar and Hage find that "size is much less likely to be related to complexity than is structural differentiation because the measures of size and complexity are different and the concepts are not the same" (p. 113).

Although it is a truism that "the higher the level of average training, the greater the differentiation by branches of knowledge and thus the greater the complexity and the fewer personnel are substitutable without extensive re-education," this accepted finding does not address the question, "Why does size not have an impact on complexity, that is, the number of different occupational specialties?" Dewar and Hage find that there is no causal relationship between an increase in size and an increase in the number of different specialties, with the exception of administration. This is based on "the greater number of interactions generated by increased size [and how that] complicates the task of administration." In terms of the relationship between technology and size, however, "new occupational specialties in production would be added only if the technology becomes more complex, and increased size does not produce this effect; the major effect of increased size is on task and not on person specialization" (pp. 113-114).

This leads the authors to argue that "in production technologies, as opposed to administrative or managerial ones, increasing size does not result in more occupational specialties either because more specialties are simply added or because existing specialized tasks can be decomposed into additional specialized parts" (p. 114). Task scope, since it informs an organization as to technological need and the variety of levels of specialization that the personnel have, should "have strong impact on the number of departments since technology is one of the major organizing principles of horizontal differentiation, although it is not for vertical" (p. 118). By following both "associations of levels and change rates it is possible to better understand how growth and increase in task scope affect both the complexity of the division of labor and structural differentiation." Whereas "specialization is associated both with growth and with increases in the number and difficulty of tasks...new activities [require] the hiring of more occupational specialties." While technological diversity requires person specialization in the production operation, both task- and person-specialization "make their impact in different areas of the organization" (pp. 129-130). Ultimately, it is the addition of new activities that drives the addition of new hires into a broader array of occupational specialties; size has "little effect on change rates in complexity," though they are positively associated with the number of hierarchical levels, whereas task scope is not (p. 130).

Individual and Organizational Contextual Variables

Kimberly and Evanisko (1981) examine "the combined effects of individual, organizational, and contextual variables on organizational adoption of two types of [technology] innovation": technology assistive to core functions and technology that supports administrative work. Their study takes into account "a large number of organizations with the objective of moving toward a more comprehensive treatment of organizational innovation that heretofore has been found in the literature" prior to 1981. Their primary goal is to "identify the relative contribution of a number of factors to an explanation of observed variability in adoption of both technological and administrative innovations by hospitals." The authors hypotheses are proposed to be "generalizable beyond [the health care] sector" (p. 691).

The authors found that "individual, organizational, and contextual variables were...much better predictors of hospital adoption of technological innovations than of administrative innovations...Organizational level variables, size in particular, were clearly the best predictors of both types of innovation" (p. 689). Their findings are the result of "a comparative analysis of the effects of variables from three different levels of analysis on organizational adoption of two different types of innovation [which were] designed to confront three issues in previous work on organizational innovation: the focus of most studies on a single innovation or class of innovations; the frequent use of sample sizes too small to permit application of multivariate analytic techniques; and the scarcity of studies examining the combined effects of individual, organizational, and contextual factors on adoption of innovation" (p. 708).

Kimberly and Evanisko come to three "primary conclusions": first, their variables were "much better predictors of the adoption of technological innovations than of administrative innovations. [Second], adoption of the two different types of innovations was not influenced by identical sets of variables. Analysis of the separate effects of variables from the three levels of analysis revealed that only one variable from each level was a significant predictor of adoption for both types of innovation. The educational level of the hospital administrator, the size of the organization, and the presence of competition in the local environment were significant predictors of both technological and administrative innovation in the separate analyses. In the analysis of their combined effects, only size was a significant predictor of both types." Finally, "organizational level variables—and size in particular—are indisputably better predictors of both types of innovation than either individual or contextual level variables. In the case of technological innovations, the only non-organizational level variable that emerged as a significant predictor was the age of the hospital, which had been conceptualized as a contextual variable. And in the case of administrative innovation, the only significant non-organizational level predictor was the cosmopolitanism of the hospital administrator, although the educational level of the hospital administrator and the age of the hospital approached" a significant level (pp. 708-709).

Sectoral Patterns of Change

Pavitt (1984) begins with the accepted premise that "the production, adoption and spread of technical innovations are essential factors in economic development and social change, and that technical innovation is a distinguishing feature of the products and industries where high wage countries compete successfully on world markets." His paper explains sectoral patterns of technical change that derive from his study of data regarding "2000 significant innovations in Britain since 1945." This data provides technological knowledge—information that is not "generally applicable and easily reproducible, but specific to firms and applications, cumulative in development and varied amongst sectors in source and direction" (p. 343).

Pavitt studied "innovating firms principally in electronics and chemicals, [which] are relatively big, and...develop innovations over a wide range of specific product groups within their principal sector, but relatively few outside." He juxtaposed these firms with "mechanical and instrument engineering [which] are relatively small and specialized, and...exist in symbiosis with large firms, in scale intensive sectors like metal manufacture and vehicles, who make a significant contribution to their own process technology." However, he also took into consideration manufacturing firms whose "process innovations come from suppliers," and from these data derived a "three-part taxonomy...(1) supplier dominated; (2) production intensive; (3) science based" to describe how "sources of technology, requirements of users, and possibilities for appropriation [affect] our understanding of the sources and directions of technical change, firms' diversification behavior, the dynamic relationship between technology and industrial structure, and the formation of technological skills and advantages at the level of the firm, the region and the country" (p. 343).

The benefits of determining "sectoral patterns of technical change" include facility with exposing "similarities and differences amongst sectors in the sources, nature and impact of innovations, defined by the sources of knowledge inputs, by the size and principal lines of activity of innovating firms, and by the sectors of innovations' production and main use" (p. 343). The importance of "building systematically a body of knowledge—both data and theory that both encompasses the production of technology, and reflects sectoral diversity" is its contribution to providing an accurate conceptualization of technical change as it occurs in "a modern economy" (p. 370).

Substitution Forecasting

Fisher and Fry (1971) advance "a substitution model of technological change based upon a simple set of assumptions" that are premised by the absence, in all circumstances, of "unconscious or invisible tampering by the forecaster in his efforts to make the future what he wants it to be." Ease of applicability and interpretation are the goals of their forecasting model for technological change, one which sees "advancing technology as a set of substitution processes [that] may seem evolutionary or revolutionary, depending upon the time scale of the substitution." In all cases, the goal of technological change is to promote the ability of users to "perform an existing function or satisfy an ongoing need differently from before. The function or need rarely undergoes radical change. Whenever exceptions to this view are found, the notion of competitive substitution as a model for technological change does not apply" (pp. 88, 75).

New methods of technological development, when first introduced, have "greater potential for improvement and for reduction in cost" because the new methods are not encumbered by the "entrenched processes of the older methodologies with which they're competing" (pp. 75-76). Their substitution model "can prove useful to a number of types of investigations, such as: forecasting technological opportunities, recognizing the onset of technologically based catastrophes, investigating the similarities and differences in innovative change in various economic sectors, investigating the rate of technical change in different countries and different cultures, and investigating the limiting features to technological change" (p. 88).

QUALITATIVE PERSPECTIVES ON TECHNOLOGY DEVELOPMENT AND INNOVATIVE PRACTICE

Transilience

Abernathy and Clark (1985) contribute a new descriptive framework that, despite the metaphoric quality of the article's title, offers researchers a concrete classification schema of four different approaches to product development and marketplace innovation. While most firms will encounter circumstances in which each of the four sub-types of "transilience," the capacity of an innovation to influence a firm's existing resources, skills, and knowledge—its established systems of production and marketing—will come into play, the authors present their transilience framework as a way of understanding that "innovation is not a unified phenomenon: some innovations disrupt, destroy and make obsolete established competence; others refine and improve" (pp. 3-5).

Transilience is the umbrella term that subsumes four types of innovation: radical, architectural, niche creation, and revolutionary. Each of these categories is aligned with "different patterns of evolution and to different managerial environments." The authors describe "the role of incremental technical change in shaping competition and on the possibilities for a technology based reversal in the process of industrial maturity," a process they term "de-maturity" (p. 3). For Abernathy and Clark, the employment of any type of innovative practice is based in a premise: "that competitive advantage depends on the acquisition or development of particular skills, relationships and resources." The significance of that competitive advantage derives from the ways in which innovation affects that development (p. 4).

Although technological innovation results in some kind of change, that change is not necessarily destructive. In product technology, for example, innovation might improve product design or make the distribution pathways more attractive to new markets or improve delivery to existing ones. Process technology innovation "may require new procedures in handling information, but utilize existing labor skills in a more effective way." These are examples of how innovation instigates change but also conserves "the established competence of the firm, and if the enhancement or refinement is considerable, may actually entrench those skills, making it more difficult for alternative resources or skills to achieve an advantage." Looking from the opposite direction, this kind of process and product innovation—beneficial to a firm's operation—may also "have an effect on competition by raising barriers to entry, reducing the threat of substitute products, and making competing technologies (and perhaps firms) less attractive" (pp. 6-7).

Transilience is composed of four types of innovation, each considered a quadrant of the concept. Radical innovation, "instead of enhancing and strengthening...disrupts and destroys. It changes the technology of process or product in a way that imposes requirements that the existing resources, skills and knowledge satisfy poorly or not at all. The effect is thus to reduce the value of existing competence, and in the extreme case, to render it obsolete." Architectural innovation "defines the basic configuration of product and process, and establishes the technical and marketing agendas that will guide subsequent development. In effect, it lays down the architecture of the industry, the broad framework within which competition will occur and develop...Using new concepts in technology to forge new market linkages is the essence of architectural innovation." Niche creation innovation serves to "conserve and strengthen established designs [by building] on established technical competence, and [improving] its applicability in the emerging market segments" when a firm seeks to open "new market opportunities through the use of existing technology" (pp. 7, 10). Opposing niche creation innovation, in which the "creation of niches and the laying down of a new architecture involve innovation that is visible and after the fact apparently logical…'Regular' innovation is often almost invisible [and] involves change that builds on established technical and production competence and that is applied to existing markets and customers." Revolutionary innovation—that which "disrupts and renders established technical and production competence obsolete, yet is applied to existing markets and customers, is the fourth element of transilience" (p. 12).

Four types of innovation comprise a "transilience map [that] is thus much more than a simple categorization of technical change; it provides a framework within which one can examine the relationships among innovation, competition and the evolution of industries, as well as develop insight about the strategies of specific competitors." Moreover, each type of innovation "tends to be associated with a different competitive environment" (pp. 14, 20).

Abernathy and Clark also discuss the industrial environment preconditions for "de-maturity, or technological innovation resulting in a reversal of an industry's maturity." There are "new technical options that open up possibilities in performance or new applications that the existing design concepts could meet only with great difficulty or not at all. These options may come through research and development from within the industry, or they may be the basis for an invasion by competitors from a related field." Second, they may come from "changes in customer demands [that] may impose requirements that can best be met with new design approaches. Third is "government policy. Regulations imposed on an established industry…may set technical requirements or demand performance standards that favor revolutionary or architectural strategic development" (p. 18). The authors conclude by suggesting that innovative firms, despite a "dominant orientation," should expect to confront the challenges of reacting to market and R&D forces that compel instituting multiple strategies based in the translience map, as innovation leads to new ways to promote regular development as well as new product design and forays into new niches (p. 21).

Evolution and a Cyclical Model of Change

Citing a paucity of research "on the nature and dynamics of technological change," Anderson and Tushman (1990) report on the development and testing of "a cyclical model of technological change" which has affinities to basic concepts of evolution and Khunian paradigm shifts (p. 604). In the authors' model, much like the Khunian model, "a technological breakthrough, or discontinuity, initiates an era of intense technical variation and selection, culminating in a single dominant design. This era of ferment is followed by a period of incremental technical progress, which may be broken by a subsequent technological discontinuity." Anderson and Tushman examine technological discontinuity as it pertains to established industries such as cement and glass, as well as a relative newcomer, the minicomputer, to find that "when patents are not a significant factor, a technological discontinuity is generally followed by a single standard [and] sales always peak after a dominant design emerges. Discontinuities never become dominant designs, and dominant designs lag behind an industry's technical frontier. Both the length of the era of ferment and the type of firm inaugurating a standard are contingent on how the discontinuity affects existing competences. Eras of ferment account for the majority of observed technical progress across these three industries" in their study. Although prior research indicates that "the core technology of an industry evolves through long periods of incremental change punctuated by technological discontinuities," the authors suggest that "a breakthrough innovation inaugurates an era of ferment in which competition among variations of the original breakthrough culminates in the selection of a single dominant configuration of the new technology" (p. 606).

Citing Tushman and Anderson's 1986 definition of technological discontinuity: "an order-of-magnitude improvement in the maximum achievable price vs. performance frontier of an industry [in which] technological change...consists of long periods of incremental change punctuated by discontinuities," the authors present a modification that stresses the radical events that comprise technological discontinuity (p. 607). "A technological discontinuity is identified when an innovation (a) pushes forward the performance frontier along the parameter of interest by a significant amount and (b) does so by changing the product or process design, as opposed to merely enlarging the scale of existing designs" (p. 620). Although the competition between older and newer technologies is often raucous, the disparagement of new technologies is based on the fact that "they frequently do not work well and are based on unproven assumptions and on competence that is inconsistent with the established technological order." This prompts in the existing "community of practitioners [an] increase [in] the innovativeness and efficiency of the existing technological order" (p. 611).

Challenging the status quo in any environment is fraught with instability and unintended consequences. Once a dominant design—a "single architecture that establishes dominance in a product class"—is on the scene, "future technological progress consists of incremental improvements elaborating the standard[,] and the technological regime becomes more orderly as one design becomes its standard expression" (p. 613). In the authors' view, dominant designs, not surprisingly, reduce variation, uncertainty, and the ease with which a product class's dominant design can be upset. In times of technological discontinuity, when new technologies are being introduced and standards challenged, the choice of a variant on the dominant model brings with it a customer incurring the expense of change or giving up the benefits "of adopting a standard, which typically include scale economies, access to an infrastructure designed around the standard, and so forth" (pp. 614-615).

After indicating how "technological progress is driven by numerous incremental innovations [in the form of] elaborating the retained dominant design, not challenging the industry standard with new, rival architectures...moving the focus of competition...from higher performance to lower cost and to differentiation via minor design variations and strategic positioning tactics," the authors suggest that "social structures arise that reinforce this stable state; standard operating procedures are predicated on the reigning technical order, organizational power structures reflect dependencies that are partly governed by technology, and institutional networks with powerful norms arise whose shape is partly determined by an industry's technical regime" (p. 618). Technological evolution has organizational consequences, including the demand to "develop diverse competences both to shape and deal with technological evolution...either to imitate these discontinuities or respond rapidly [and] combine technological capabilities with the ability to shape inter-organizational networks and coalitions to influence the development of industry standards" (p. 629).

Technologies, Organizational Structure, and Historical Processes

Barley's 1986 *Administrative Quarterly* article presents a sociologically based argument intent on explaining the connection between "institution and action to outline a theory of how technology might occasion different organizational structures by altering institutionalized roles and patterns of interaction." When viewed through a social science lens, the fact that technology shapes the structure of an organization is a fundamental truism. Social scientists provide substantial evidence that "technologies transform societies by altering customary modes and relations of production." Barley's 1986 article uses data drawn from a study of how one technology—a CT scanner—used in two different settings, changes "the organizational and occupational structure of radiological work." Instead of presenting the data through traditional management theories of organizational form, theories that Barley finds to be "insensitive to the potential number of structural variations implicit in role-based change," he turns to sociology to "understand how technologies alter organizational structures," suggesting that "researchers may need to integrate the study of social action and the study of social form" (p. 78).

The standard sense of the term "structure" for those studying technology and organization refers to "abstract, formal relations that constrain day-to-day action in social settings. In fact, when we view structure as a formal constraint, "three other presumptions have typically followed: that technology is a material cause;

that relations between technology and structure are orderly; and that these relations hold regardless of context" (p. 79).

An alternative view on structure provided by organizational theorists promotes the idea of its being "patterned action, interaction, behavior, and cognition." Barley finds neither view sufficient, calling for a conflation or synthesis of both, suggesting that structure is "both a product of and a constraint on human endeavor." Barley bases this call for synthesis in *negotiated-order theory* (Strauss, 1978, 1982), a derivative of symbolic interactionism which examines quotidian events, and *structuration theory,* which combines functionalist and phenomenological views of social order (Giddens, 1976, 1979). Notwithstanding significant differences among these approaches, Barley (1986) subscribes to their common denominator, that "adequate theories must treat structure as both process and form" (p. 79).

We can consider structure to consist of rules that limit conduct and modify action; Barley uses language as an analogy for understanding structure. Both structuration and negotiated-order theory "attempt to bridge the gap between a deterministic, objective, and static notion of structure, on one hand, and its voluntaristic, subjective, and dynamic alternative, on the other, by positing two realms of social order (analogous to grammar and speech) and by shifting attention to the processes that bind the two together." They both view structure as the product of actors' interpretations of events, the nature and scope of access to resources, and ethical frameworks that shape the social order. Therefore, when we study organizational structure, we are actually "investigating how the institutional realm and the realm of action configure each other" (p. 80).

Researchers need to keep in mind that even though these patterns become engrained through repetition and reinforcement, there are events that alter them, and these alterations change the institutional structure of an organization, ultimately becoming accepted as the status quo. We may even consider structural changes from an evolutionary perspective, "since technologies occasion adaptations whose implications may congeal but slowly as actors redefine their situation" (p. 81). Barley calls on researchers to document "traditional patterns of behavior, interaction, and interpretation before the technology arrives. Such assessment is critical not only because institutional patterns influence the action that surrounds the technology's adoption, but also because such patterns set contextually specific baselines for judging structural stability and change. Once the technology arrives, attention shifts from the institutional context to the social practices that envelop the technology's use, in order to document behaviors and cognitions, which are the raw material from which interaction orders emerge" (p. 83). In the end, "decision makers may in fact influence the evolution of interaction orders, but the structural consequences of their decisions are likely to be unanticipated. Structuring theory thus departs from previous approaches to the study of technology by postulating that technologies are social objects capable of triggering dynamics whose unintended and unanticipated consequences may nevertheless follow a contextual logic. Technologies do influence organizational structures in orderly ways, but their influence depends on the specific historical process in which they are embedded. To predict a technology's ramifications for an organization's structure therefore requires a methodology and a conception of technical change open to the construction of grounded, population-specific theories" (p. 107).

A Role-Based Approach to Technology and Innovation

Barley's 1990 *Administrative Science Quarterly* article looks at the relationships between social actors and technology, and their effects on subsequent action. By viewing technological development through a sociologist's lens, Barley attempts to "overcome four shortcomings that characterize much previous research on technology and structure: ambiguous terminology, reliance on distant knowledge, inferential leaps between levels of analysis, and the use of nonsocial concepts" (p. 64). This research indicates that "technologies change organizational and occupational structures by transforming patterns of action and interaction." Barley offers a different perspective—"the microsocial dynamics occasioned by new technologies reverberate up levels of analysis in an orderly manner [and] a technology's material attributes are said to have an immediate impact on the non-relational elements of one or more work roles. These changes, in turn, influence the role's relational elements, which eventually affect the structure of an organization's social networks. Consequently, roles and social networks are held to mediate a technology's structural effects" (p. 61).

Whether one defines technology as "apparatus, machines, and other physical devices [or] technique, the behaviors and cognitions that compose an instrumental act [or even] in the sense of organization, a specific arrangement of persons, materials, and tasks," technology is developed and used in social "structures" that "delineate a hierarchy of increasing abstraction or aggregation" manifested in "the repetitive features of day-to-day activity, the formal attributes of organizations, and even more global institutional arrangements such as the bureaucratic ideal or professional dominance" (p. 65). Barley suggests that it is best to consider roles "as bundles of non-relational and relational elements that can be separated only analytically" (p. 68). Non-relational elements are defined as "the set of recurrent activities that fall within the purview of a person who assumes a particular position or job [and] encompass all the behaviors that individual's ordinarily perform as role incumbents, regardless of whether the behaviors are construed as obligations or are explicitly sanctioned. Because non-relational elements of a role include skills and tasks, it is here that technologies are likely to have their most immediate impact"

(p. 69). The benefit of the "role-based approach [is that it] explicitly articulates how skills, tasks, and activities influence role relations and how role relations, in turn, affect an organization's and occupation's structure" (p. 98).

Innovation, Design, and Choice

Like Barley, Clark (1985) employs an evolutionary perspective to advance a conceptual framework "for analyzing the sequence of technological changes that underlie the development of industries," particularly "the interaction between technical innovation and customer demands" (p. 235). Clark is interested in presenting "a detailed description of the forces shaping the pattern of innovation that emerges from this process [by examining] both the decisions of producers in the design of products (and processes) and choices of customers" and concentrating on "the sequence of design decisions that emerge over time" (pp. 236-237)

Discussion of technological change and development incorporates uncertainty, search behavior, and learning as important factors that impact on the organizational and managerial practices of technological firms. This is because there is a plethora of technological choice and a dearth of awareness regarding "customer needs or the link between technology and preferences." Considering that "innovation is relatively rapid, and fundamental [and that the] production process in turn, must be highly flexible, relatively labor intensive, and somewhat erratic in work flow," the notion that products and processes evolve from "an early, 'fluid' state, to one that is highly 'specific' and rigid" is not unreasonable. The concept of uncertainty highlights not only that it "is more than a precondition for evolution, it is also a determinant of its pattern" (pp. 235-236).

Clark also highlights the concept of "technical interdependence," which he defines as "the tendency for designs to build on one another, and the role of perceived technical opportunities in determining the pattern of technical advance." Accepting that innovation is shaped by technical interrelationships and interdependencies, he points out that "interaction with market demands is likely to be important as well": not only will feedback from the market "influence technical development, but...design choices may also influence the evolution of concepts that guide customer choice." Ultimately, "uncertainty about technology and customer preferences leads to a diversity of technology in the products vying for customer acceptance" (pp. 237-238).

For Clark, discernable patterns of innovation derive from "the logic of problem solving in design [and] the formation of concepts that underlies customer choice, [imposing] a hierarchical structure on the evolution of technology." This is a result of design being "a search for understanding of what the object or product is, and [therefore] ought to be given the context in which it must function." Oftentimes, new products, presenting customers "with a set of unfamiliar possibilities," create a "problem of choice [involving] both the formation of concepts with which to understand the product, and the development of criteria to be used in evaluation." Clark promotes the idea "that the success of an innovation will be influenced by...users' 'need determinateness,' the extent to which preferences are specified (or need satisfaction is expressed) in terms of product classes, functions and features" (pp. 241, 244). "Ultimately, like product development, process development, guided by the logic of problem solving and the development of customer concepts, arises out of a search for solutions to problems of design" (p. 247).

Technological Innovation and the Scientific Paradigm

Apparent from its title, Dosi's 1982 *Research Policy* article attempts to draw analogical parallels between the Khunian sense of scientific "paradigm," essential to the concepts of methods and goals in scientific theory, and the domain of technological innovation. "Technological paradigms and trajectories, are in some respects metaphors of the interplay between continuity and ruptures in the process of incorporation of knowledge and technology into industrial growth: the metaphor, however, should help to illuminate its various aspects and actors and to suggest a multi-variables approach to the theory of innovation and technical change" (p. 161). Dosi's "model tries to account for both continuous changes and discontinuities in technological innovation. Continuous changes are often related to progress along a technological trajectory defined by a technological paradigm, while discontinuities are associated with the emergence of a new paradigm." This is in response to what Dosi considers to be inadequate—the "one-directional explanations of the innovative process, and in particular those assuming 'the market' as the prime mover [as explanations of] the emergence of new technological paradigms." Researchers must keep in mind that the "history of a technology is contextual to the history of the industrial structures associated with that technology. The emergence of a new paradigm is often related to new 'Schumpeterian' companies, while its establishment often shows also a process of oligopolistic stabilization" (p. 147).

For Dosi, "a 'technological paradigm' [is] broadly in accordance with the epistemological definition as an 'outlook', a set of procedures, and a definition of the 'relevant' problems and of the specific knowledge related to their solution." As such, "technological paradigm" is a term that depicts "its own concept of 'progress' based on its specific technological and economic trade-offs." Dosi is concerned with offering a model that will allow researchers to understand and predict the direction of technological advances, and to be able to do so through analysis of "the role played by economic and institutional factors in the selection and establishment of those technological paradigms and the interplay between endogenous economic mechanisms and technological innovations." By doing so, researchers can advance beyond the standard duo of theoretical approaches: demand-pull, in which "market forces [serve] as the main determinants of technical change," and technology-push, which sees technology "as an autonomous or quasi-autonomous actor, at least in the short run" (p. 148).

If one views the market demand-pull theory as a linear process or series of events, it would consist of five segments: (1) Existing goods on the market satisfy (2) consumers who "express their preferences about the features of the goods they desire (i.e., the features that fulfill their needs the most) through their patterns of demand, different 'needs' by the purchasers." (3) Market demand then grows in proportion to consumer income, resulting in "proportionally more of the goods which embodied some relatively preferred characteristics," causing (4) producers to act by adjusting prices to demand, and thereby stimulating the innovative process, in which (5) "successful firms will at the end bring to the market their new/improved goods, letting again the 'market'...monitor their increased capability to fulfill consumers' needs" (p. 149). One of the failures of the demand-pull approach, for Dosi, is its inability "to produce sufficient evidence that needs expressed through market signaling are the prime movers of innovative activity" (p. 150). Both approaches, demand-pull and technology-push, are considered simplistic and inadequate, as they fail to take into account the notion that new "technologies are selected through a complex interaction between some fundamental economic factors...together with powerful institutional factors...Technical change along established technological paths, on the contrary, becomes more endogenous to the 'normal' economic mechanism" (p. 157).

CONCLUSION

The consideration of technology as an explanatory element in organizational analysis (Hage & Aiken, 1969) is foundation for much of the later research in organizations and innovation. Aldrich (1972) found that technology is best treated as an independent variable when considered as part of a group of organizational variables.[1] The literature on technology and size were found to be closely related to the two variables: structural differentiation and complexity (Dewar & Hage, 1978). The scope of an organization's task is more important to the degree of labor differentiation, as opposed to an organization's size. Size appears to be related to structural differentiation, but the relationship between size and complexity was found to be not as evident.

Important foundational work has also considered innovation and organization from the perspective of the interaction between the firm and innovation characteristics. By studying the adoption of both technological and administrative innovations in a hospital, Kimberley and Evanisko (1981) are able to determine that individual, organizational, and contextual variables are better predictors of hospital adoption of technological innovations than of administrative innovations. However, organizational-level variables, especially size, are the best predictors of both types—administrative and technical innovation. Challenges to attempts to further develop an integrative theory of innovation adoption and organization adoption are suggested to rest heavily on the identification of four sources of instability in empirical results. The sources are: (1) variation among primary attributes, (2) interaction, (3) ecological inferences, and (4) varying modes of operationalization of innovation (Downs & Mohr, 1976). Considering the effects of the sources of instability on firms and innovation under consideration is critical for empirical work that considers technology innovation management from an organizational perspective. Recognition that most innovation does not result in things that are completely new, but rather an act of substitution from the existing entity to an innovation that has one or more preferable characteristics, resulted in a widely used model to forecast adoption of innovation that involves substitution of an existing product for the now innovative product that is being offered (Fisher & Fry, 1971).

Critical foundational findings are not only found by considering innovation from the perspective of the organization. Due to the wide variation in success when working with innovation, however one defines success, Nelson and Winter (1977) describe our knowledge and research in the area as *having a number of distinct intellectual tradeoffs*. They also point out differences in the management and support of innovation in different sectors. To address these problems they offer theoretical structure to bridge subfields of the study of innovation and to treat uncertainty and institutional diversity centrally. The success of such an approach is demonstrated by Pavitt (1984), who studied sectoral patterns of technical change and determined the presence of three categories of innovation: (1) supplier dominated, (2) production intensive, and (3) science based. His taxonomy assists in understanding why certain types of innovation are more likely to occur in certain locations and circumstances.

Through the pair-wise comparison of successful and unsuccessful innovations (Rothwell et al., 1974), successful innovators were found to:

1. Have a much better understanding of user needs;
2. Pay more attention to marketing and publicity;
3. Perform their development work more efficiently, but not necessarily more quickly;

4. Use outside technology and scientific advice, not necessarily in general but in the specific area concerned; and
5. Are more senior and have greater authority than their counterparts who fail.

In addition to the insights that are offered through empirical assessment, researchers also inquire into the nature of innovation and its impact and implications on firms from a qualitative perspective. Dosi (1982) provides a model, which through reference to Kuhnian theory works to bring together different types of innovation, technology trajectories, and paradigm shifts. The relationship between new paradigms and new 'Schumpeterian' companies is noted. The "transilience" framework further develops the notion of different types of innovation that firms may experience (Abernathy & Clark, 1985). The framework offers four types of innovation: radical, architectural, niche creation, and revolutionary. The integration of Khunian paradigm shifts with a model of cyclical technological change is further developed by Anderson and Tushman (1990). As in Kuhn's work, a technological breakthrough results in an *era of ferment* involving intense technical variation and selection. The result is a single dominant design, which is followed by incremental technical progress until the next technological breakthrough.

Finally, the limitations of considering innovation's effects on organizations, using organization theory, are shown to be the insensitivity of organization theory to the number of structural variations implicit in role-based change (Barley, 1986). By considering innovation from the perspective of a sociologist, it was shown how technologies may alter organizational structures. The sociologist's perspective is applied once again and finds that shortcomings of earlier research on technology and structure were associated to ambiguous terminology, reliance on distant knowledge, inferential leaps between levels of analysis, and the use of nonsocial concepts (Barley, 1990). Technology affects roles at all levels of analysis, and these changes affect the relational aspects of a role. Consequently, roles and social networks mediate a technology's structural effects. Alas, there is a need not only to consider the interaction of innovation with members, stakeholders, and employees of the supply chain, but also to consider customer demands and how the interaction of customer, supplier, and technology affect the development of product. Clark (1985) demonstrates how these insights can be obtained through the consideration of the evolution of design decisions.

REFERENCES

Abernathy, W.J., & Clark, K.B. (1985). Innovation: Mapping the winds of creative destruction. *Research Policy, 14,* 3-22.

Aldrich, H.E. (1972). Technology and organizational structure: A re-examination of the findings of the Aston Group. *Administrative Science Quarterly, 17*(1), 26-43.

Anderson, P., & Tushman, M. (1990). Technological discontinuities and dominant designs: A cyclical model to technological change. *Administrative Science Quarterly, 35*(6), 604-633.

Barley, S.R. (1986). Technology as an occasion for structuring—evidence from observations of CD scanners and the social order of radiology departments. *Administrative Science Quarterly, 31*(1), 78-108.

Barley, S.R. (1990). The alignment of technology and structure through roles and networks. *Administrative Science Quarterly, 35*(1), 61-103.

Clark, K.B. (1985). The interaction of design hierarchies and market concepts in technological evolution. *Research Policy, 14*(6), 235-251.

Dewar, R., & Hage, J. (1978). Size, technology, complexity, and structural differentiation—toward a theoretical synthesis. *Administrative Science Quarterly, 23*(1), 111-136.

Dosi, G. (1982). Technological paradigms and technological trajectories—a suggested interpretation of the determinants and directions of technical change. *Research Policy, 11*(4), 147-162.

Downs, G.W., & Mohr, L.B. (1976). Conceptual issues in the study of innovation. *Administrative Science Quarterly, 21*(4), 700-714.

Fisher, J.C., & Fry, R.H. (1971). A simple substitution model of technological change. *Technological Forecasting and Social Change, 3*(1), 75-83.

Giddens, A. (1979). *Central problems in social theory.* Berkeley, CA: University of California Press.

Giddens, A. (1976). *New rules of sociological method.* London: Hutchinson.

Hage, J., & Aiken, M. (1969). Routine technology, social structure and organization goals. *Administrative Science Quarterly, 14*(3), 366-376.

Kimberly, J.R., & Evanisko, M.J. (1981). Organizational innovation—the influence of individual, organizational and contextual factors on hospital adoption of technological and administrative innovations. *Academy of Management Journal, 24*(4), 689-713.

Nelson, R.R., & Winter, S.G. (1977). In search of useful theory of innovation. *Research Policy, 6*(1), 36-76.

Pavitt, K. (1984). Sectoral patterns of technical change: Towards a taxonomy and a theory. *Research Policy, 13*(3), 343-373.

Rothwell, R., Freeman, C., Horlsey, A., Jervis, V.T.P., Robertson, A.B., & Townsend, J. (1974). SAPPHO updated—Project SAPPHO phase 2. *Research Policy, 3*(3), 258-291.

Strauss, A. (1982). Interorganizational negotiations. *Urban Life, 11,* 350-367.

Strauss, A. (1978). *Negotiations.* San Francisco: Jossey-Bass.

ENDNOTE

[1] In addition to this useful finding, this work demonstrates how path analysis can be used as a technique to consider whether a variable should be considered as dependent or independent—an important methodological question that is often overlooked.

Chapter IV
Social Influence and Human Interaction with Technology

INTRODUCTION

This chapter discusses how information that supports innovation flows throughout an organization, the construction and effects of team composition, the innovative processes that teams employ, and the development, implementation, and evaluation of systems used to manage the flow and distribution of information. As Allen and Cohen (1969) point out, effective communicators rise in their organizations as a result of their willingness to engage information—by reading and conversing outside of their immediate settings, but as Tushman (1977) explains, that kind of outreach precipitates special boundary roles, which come about to satisfy an organization's communication network's role of bridging an internal information network to external sources of information. Thompson (1965) investigates the conditions necessary to move an organization from a single-minded focus on productivity to one of those that facilitate innovation. At times, that means engaging rival firms, and von Hippel (1987) demonstrates that information sharing is economically beneficial to the organizations doing the trading. Freeman's (1991) finding that information regarding innovative processes entails the development of effective information networks confirms how important it is for successful innovation that there exist effective external and internal communication networks, and that individuals collaborate to share information. von Hippel (1994) returns later in the chapter to qualify this point by showing that there is a direct correlation between the level of stickiness and the expense related to moving that information to a location where it can be applied to solving a problem.

Bantel and Jackson (1989) begin the section on team composition by suggesting that certain demographic factors affect a team's ability to be innovative, but resource diversity—including communication ability—is ultimately essential to innovation. For Howell and Higgins (1990), identifying a champion among a team's members will facilitate innovation, while Anconia and Caldwell (1992) find that the greater the functional diversity, the more team members communicated outside of their teams' boundaries. Scott and Bruce (1994) take a different vantage point, focusing on the individual and his or her influence on and adaptation to an organization's climate for innovation. The section on innovation process begins with Hage and Dewar (1973), who conclude that ultimately, the values held by an organization's elite group are more significant when predicting innovation than the values of any single leader or even the entire staff, but the correlation between a single leader and innovation should not be dismissed as a valid predictor of an organization's ability to innovate. Even so, as Daft (1978) has found, there is evidence to support the theory that there can be opposing innovative processes in an organization: one that begins at the lower levels of its hierarchy, and one that percolates down from upper levels. Even more radical, Quinn (1985) proposes that corporate executives understand and adapt to the fact that the innovation environment is filled with surprise, characterized by chaos, and virtually immune to control. The chapter concludes with Porter and Millar's (1985) article describing how information technologies affect management strategies and how these strategies are disseminated throughout a firm.

FLOW OF INFORMATION

In 1969, Allen and Cohen set out to explain the course that scientific and technological information takes in research centers and laboratories. Are there distinct pathways that information travels as it moves from external sources to people working with research labs? Their study, premised on the idea that research done while excluding outside information into the lab will ultimately fall short, consisted of examining patterns of technical communication in two different research labs, each of which had identifiable technical communication networks that arose from the nature of social interaction and work structure. The most effective communicators, those the authors refer to as the "sociometric stars," rose to prominence in the lab environment through their willingness to "either make greater use of individuals outside the organization or read the literature more than other members of the laboratory" (p. 12). In other words, either all members of an R&D team proactively seek the latest information regarding recent developments in the field, or managers bring in knowledgeable people to serve as consultants to the staff.

They also found evidence in existing literature that the lesser the rate of internal communication in an organization, the more chance there is that the research team will perform poorly, suggesting to managers that there is a need for external sources of information. Managers and project leaders can stave off the need for external consultants by recognizing that there are different rates of information flow when one considers the different demands spurred by organizational loyalty and structure, and the value of shared experience as opposed to the organizational schema that are inculcated in an academic setting. In other words, the affiliations, loyalties, and social relationships that develop among team members can be thought of as one way to see how the world works; conversely, training and information received in an academic setting and then brought into a research facility can be thought of as a different coding scheme that "introduces the possibility of mismatch and attendant difficulties in communication between organizations" (p. 12). Having team members who understand both schema, and therefore serve as translators, can mitigate the negative effects of such a mismatch.

Allen and Cohen identify previous literature concerning the effect of "prestige or status hierarchies in a social system [on] the flow of information." Essentially, those of high status will be more sociable and therefore more communicative with one another, whereas those of lower status tend not to like one another or communicate effectively with one another, "direct[ing] most of their communication toward the higher-status members, without complete reciprocation" (p. 16). At the time of their writing, Allen and Cohen believed that gatekeepers—those who could translate and facilitate communication among different hierarchical levels of an organization—would continue to transmit information, but management of the laboratories studied failed to see the value in that role, and therefore either discourage[ed] this activity by failing to reward it, or to reward the gatekeeper by promotion and thereby [made] it impossible for him to continue as a transmitter of information" (p. 19).

Boundary Roles and Innovation

Tushman (1977) uses previous organizational behavior and research and development literature to inform his understanding of how "special boundary roles" serve as a way for innovative organizations to facilitate necessary "cross-boundary communication" in a research and development lab. Particular attention is paid to the "distribution of these special boundary roles within the organization and their impact on subunit performance," and how boundary roles factor into the innovative process (p. 587).

There is often a strong need for an innovating organization to provide information to a variety of external information areas. This need precipitates special boundary roles, which come about to satisfy an organization's communication network's role of bridging an internal information network to external sources of information. These boundary roles occur at several places in the organizational structure, and the nature of their distribution depends on the type of work occurring within the organization. Tushman's findings are consonant with other research "on boundary spanning in general and highlights the importance of boundary roles in the process of innovation" (p. 587).

Tushman summarizes previous literature on the process of innovation development and dissemination, identifying the variety of steps and phases that result in decisions and coordinative efforts, and by communication patterns. Ultimately, however, Tushman focuses on the three-step innovation process offered by Myers and Marquis (1969): "idea development (the generation of a design concept), problem solving (technical efforts and problem solving in developing the proposed idea), and implementation (pilot production, inter-area coordination)" and point out that among the various positions and descriptions of these communicative patterns, "one important difference is the locus of critical information and feedback" (Tushman, 1977, p. 588).

Although Tushman recognizes the importance "of extra-organizational communication since the laboratory must receive up-to-date information about market and technological developments," he also points to other researchers' findings regarding the relevance of understanding external information such as user or market need, as well as maintaining an awareness of trends related to "new technological products, processes, and knowledge" (p. 589).

Special boundary roles are crucial to the flow of technical information between R&D labs and the larger organizations of which they are a part. The success of new projects is due, in large part, from an effective interaction between R&D teams and units such as sales, marketing, and the factory floor. Understanding that there are blocks to information flow and the hazards that complicate "transferring information across organizational interfaces," Tushman looks to systems theory research for reasons why such problems develop. The more complex any organization becomes, the more subgroups differentiate not only their tasks and goals, but their methods of accomplishing them and the social norms and behaviors that define them. Problems arise when there are discordant norms and coding schemes. When there is an overt disconnect between subgroups, and information cannot flow across organizational boundaries, yet project success is dependent upon effective communication, it is crucial that organizations "develop special boundary roles" (p. 590).

Referring to Allen and Cohen's (1969) point that the identification of key nodes in a communication network serve as a conduit for relevant external information into an internal communication system, "communication stars" must be identified and empowered to connect external sources of information to the R&D lab, as these stars are capable of translating and applying external information to the specific needs of the project team. At the same time, Tushman raises the question of how important it is to keep the boundaries between R&D and the larger organization clearly defined, as the difficulties inherent to differing coding schemas of different subgroups, even in an R&D setting, can hinder the integration of product development. Having information flow through agents who can effectively communicate and mediate—those occupying special boundary roles—will result in progress in innovative product development. Tushman's 1977 research "indicates that special boundary roles function to link the innovating system with various sources of external information and feedback. Thus, communication with external areas is not distributed equally among the innovating unit's staff but takes place through a limited set of individuals able to translate between several coding schemes. These individuals, or special boundary roles, are well connected to external information areas and are frequently consulted within the innovating unit. These boundary roles exist to mediate communication across several organizational interfaces" (p. 602).

Productivity and Innovation

Thompson's 1965 article reminds us of the enormous efforts toward structural changes in organizations that are necessary to initiate and promote innovative thinking. Before beginning his overview of contemporary organizational design and the forces behind it, he offers one premise and one definition: "No attempt is made to answer the question as to whether innovation is desirable or not," and by "innovation is meant the generation, acceptance, and implementation of new ideas, processes, products or services. Innovation therefore implies the capacity to change or adapt" (p. 2). His approach is comparative, taking typical bureaucratic architectures and juxtaposing them to what organizational and behavioral psychologists would suggest as conditions "conducive to individual creativity" (p. 1). What are the conditions necessary to move an organization from the rigid hierarchical structures that promote productivity to a variety of levels of hierarchy to facilitate innovation?

The typical successful organization can be viewed as "high [on] productive efficiency but low [on] innovative capacity" (p. 1). However, there is no synonymous relationship between adaptive and innovative organizations: a firm can adapt to new and varied pressures but fail to come up with new ideas. The distinction made between adaptive and innovative firms does not mean there is a void in terms of innovative practices. The innovative firm is capable of putting new ideas into practice. However, in order for any organization to function effectively, a "production ideology," the set of goals, objectives, and methods that "legitimizes the coercion of the individual by the group," needs to be articulated. In a vertical organization, this would mean goals set by an owner or executive group carried out through the efforts of employees who have been hired to do specific, discrete, and non-overlapping functions (p. 2).

Thompson characterizes "this [stereotypical] organization [as] a great hierarchy of superior-subordinate relations in which the person at the top, assumed to be omniscient, gives the general order that initiates all activity [and] authority and initiation are cascaded down" the chain of command, much like the military, where "complete discipline [is] enforced from the top down to assure that these commands are faithfully obeyed...each position is narrowly defined as to duties and jurisdiction, without overlapping or duplication[, and problems] that fall outside the narrow limits of the job are referred upward until they come to a person with sufficient authority to make a decision" (p. 3). While the egalitarian and horizontal organizational structure may seem to suggest that there is value in considering "the organization as a coalition" of people, skills, and efforts to achieve agreed-upon goals, "according to the Monocratic stereotype, the organization as a moral or normative entity is the tool of an owner, not a coalition" (p. 4).

How can members of stereotypical organizations adopt innovative methods when the "extrinsic reward system, administered by the hierarchy of authority, stimulates conformity rather than innovation"? The answer is to create an environment in which each member is personally committed to the organization's goals so that the rewards for creative thinking and action are primarily intrinsic rather than material. In such an environment, the concept of success is defined by the synergy between the individual and the organization's goals rather than the completion of a specific task, often preceded by the "normal psychological state...of...anxiety" (p. 6). Success eventually becomes synonymous with conformity, but that conformity is antithetical to creativity. Therefore, to "gain the independence, freedom and security required for creativity, the normal individual has to reject this concept of success" (p. 6). The "traditional bureaucratic orientation is conservative," a condition not conducive to innovation, which is in fact perceived as "threatening." Its primary concern is "the internal distribution of power and status" resulting in "the distribution of...extrinsic rewards." Reaction to innovative ideas is first and foremost characterized as "How does it affect us?" (p. 7).

To move toward an organizational structure that promotes innovation, several basic resources are needed, including "uncommitted money [and] time, [but also the human resources of] skills and good will," which inculcate a positive and unbounded sense of an individual's "limits of his capacities, so that he has that richness of experience and self-confidence upon which creativity thrives—a professional" (pp. 10-11). This will promote a "structural looseness" in the organization that puts "less emphasis on narrow, non-duplicating, non-overlapping definitions of duties and responsibilities," defines people by their "professional type rather than the duties type," creates a more open and freer sense of communication and "decentralizes...assignment and resource decisions" (p. 13). How costly this approach would be is something that Thompson, in 1965, simply could not predict, as management theorists "do not know the value of the novel ideas, processes, and products, which might be produced by the innovative organization, and we do not know that our present methods of costing and control are the best approach to achieving low-cost production" (p. 20). He believes, though, "that bureaucratic organizations are actually evolving in this direction," and that evolution is manifesting itself in attempts toward "increased professionalization, a looser and more untidy structure, decentralization, freer communications, project organization when possible, rotation of assignments, greater reliance on group processes, attempts at continual restructuring, modification of the incentive system, and changes in many management practices" (p. 20).

Informal Trading Knowledge

In von Hippel's 1987 article, his focus is on "the informal trading of proprietary know-how between rival (and non-rival) firms." Presaging the positive, cooperative spirit of the open source movement, von Hippel suggests that such information sharing is economically beneficial to the organizations doing the trading, but also that such activity is potentially beneficial in "any situation in which individuals or organizations are involved in a competition where possession of proprietary know-how represents a form of competitive advantage" (p. 291).

Innovation and Information Networks

Freeman's 1991 article begins by identifying and summarizing a major focus of 1960s empirical research on the flow of information from external sources into innovative business organizations, resulting in a reiteration of "the vital importance of external information networks and of collaboration with users during the development of new products and processes...whilst recognizing the inherent element of technical and commercial uncertainty" (p. 499). Freeman builds on von Hippel's sense of benefit deriving from industry-wide cooperation and seeks new ideas related to the process of networking innovators and their ideas. He discusses the development, in the 1980s, of flexible, regional networks that promote cooperative research among competing organizations, "and whether they are likely to remain a characteristic of national and international innovation systems for a long time to come, or prove to be a temporary upsurge to be overtaken later by a wave of take-overs and vertical integration" (p. 499).

By identifying and comparing innovations that succeed, and juxtaposing them against those that fail, Freeman finds that the failures both had lower resource investment and resulted in poorer product development. Comparative components included the size of the firm, which did not "discriminate between success and failure" (p. 500), and the size of a distinct research and development project, which did discriminate. Information regarding the design, manufacture, and sale of new products, and the innovative processes embedded therein, entailed the essential role of information networks "both in the acquisition and in the processing of information inputs" (p. 501). This confirms how important it is for successful innovation that there exist effective external and internal communication networks, and that they collaborate to share information.

How these networks are organized affects how an organization addresses systemic innovation, as the information networks themselves are "an inter-penetrated form of market and organization" that contain differing levels of cohesiveness among their members. Freeman defines a "network" as "a closed set of selected and explicit linkages with preferential partners in a firm's space of complementary assets and market relationships, having as a major goal the reduction of static and dynamic uncertainty." Freeman indicates that cooperation among regional firms, "as a key linkage mechanism of network configurations," can take a variety of forms, including "joint ventures, licensing arrangements, management contracts, sub-contracting, production sharing and R&D collaboration" (p. 502). Allowing that "in the early formative period of any major new technology system, almost by definition there are no dominant designs or standards and a state of organizational flux" (p. 510), it is logical that innovative business groups are not weighed down or restricted by industry standard or characteristic R&D techniques, but as adoption of technologies and methodologies takes place, "economies of scale become more and more important and standardization takes place," thus reducing the number of competing firms. However, Freeman points out that there is an alternative outcome to such standardization, particularly when one includes information networks as essential components to innovation, and that is that the number and productivity of autonomous firms "will grow still more important and will become the normal way of conducting product and process development" (p. 510).

Sticky Information

von Hippel returns to this cohort of top-cited articles on information flow in his 1994 piece on "sticky information" and its role in problem solving in innovative environments. He stresses the importance of uniting information and problem-solving skills in one location, be it on site in a lab or virtually via the Internet, as a prerequisite to solving any problem in the scope of product innovation. As long as information is easily available and inexpensive or free to get and share with others, where the information needed to solve a problem is used, is not at issue; but "when information is costly to acquire, transfer, and use—is, in our terms, 'sticky'—we find that patterns in the distribution of problem solving can be affected in several significant ways" (p. 429). There is a direct correlation between the level of stickiness and expense related to moving that information to a location where it can be applied to solving a problem. The thrust of the paper is on "four patterns in the locus of innovation-related problem solving that appear related to information stickiness" (pp. 429-430). Underlying these patterns are a few basic premises. First, when sticky information that problem solvers need is located in one place, the people will gravitate toward the information. When that information is distributed, problem solvers may do portions of their work at each of the locations. The third pattern is driven by cost: when the information is particularly sticky (expensive to obtain and/or share), problem solvers will divide their goal into sub-problems that can be addressed using the information held at any individual site. Lastly, not all locations of sticky information will receive the benefit of investment intended to reduce the expense of information at that location. Issues arising from the sticky information phenomenon result in the following four patterns, according to von Hippel (1994, p. 429): "patterns in the diffusion of information, the specialization of firms, the locus of innovation, and the nature of problems selected by problem solvers."

von Hippel draws an analogy between the costs of information transfer in problem solving and those of the physical manufacture of a product. Just as when there is sticky information needed to solve a problem, drawing the activity to the information to keep expenses low, a manufacturing firm will try to locate its facilities in such a way as to keep its transportation costs as low as possible. Citing research that indicates trial and error as a significant component of problem solving in cases in which sticky information is located at multiple sites, von Hippel suggests that problem solving will "sometimes move iteratively among these sites" or efforts will be extended toward "reducing the stickiness of some of the information" (pp. 432, 436).

TEAM COMPOSITION

Information Processes and Innovation

Bantel and Jackson's 1989 paper takes a demographic approach to investigating the leadership characteristics that result in a tendency to innovate, rather than avoid new approaches to both technical (products, services, and systems) and administrative (human and organizational) innovation. Previous research in this domain approached top decision makers such as CEOs as individuals making decisions on their own. Bantel and Jackson's demographic approach focused on characteristics as variables of top management teams: "We assume this dominant coalition acts as a decision-making unit for the organization" (p. 107). The authors looked at the demographics of top decision-making teams in 199 banks to discover the elements of team composition that supported innovative activity. Innovations "were identified through reference to the state of the art in the industry" (p. 108).

"The demographic characteristics of top management teams [that] were examined [included]: average age, average tenure in the firm, education level, and heterogeneity with respect to age, tenure, educational background, and functional background. In addition, the effects of bank size, location (state of operation), and team size were assessed. Results indicate that more innovative banks are managed by more educated teams who are diverse with respect to their functional areas of expertise. These relationships remain significant when organizational size, team size, and location are controlled for" (p. 107).

There are two basic organizational theory approaches to understanding the role that leaders play in effectuating a company's performance. Some hold that leaders and their abilities are environmentally determined (and therefore have relatively little ability to control organizational structure or reshape factors that support action), while others look at leaders as proactive decision makers who have a great deal of power over the direction of a firm. However, there is a middle position, which views organizational leaders as conduits that allow external influences into their firms, "thereby facilitating adaptation to the environment" (p. 107). These perspectives have shaped organizational behavior research, resulting in yet another opposition of approaches. Bantel and Jackson divide the field into the "direct assessment approach," which "directly assess[es] the psychological attributes of decision-makers and examine[s] their relationship to outcomes," and the "demographic approach," which the authors view as being more practical, but in terms of statistical analysis, has the disadvantage that "characteristics do not co-vary perfectly with the psychological attributes of interest" (pp. 107-108). These different approaches suggest differing hypotheses to study, with the psychological approach stressing "the role of cognitive resources in group problem solving," while the demographic approach suggests significant value in "the role of cohort effects in organization processes" (p. 108).

Perhaps more fundamental than the divergence of research approaches, the authors point out that there is not agreement on what "innovation" means. They describe the term as being actively inflected (innovation as process), as nominally inflected (innovation as the end product, a program or a service), and adjectivally (innovation as an descriptive attribute that an organization manifests). Their review of the literature indicates that most references to the term often take the nominal sense—the thing the firm sells. "In other words, the innovation 'process' culminates with innovation 'items', and firms that cycle through the process relatively frequently are described as 'innovative'" (p. 108).

The authors' analyses resulted in identifying "three phases of the decision process: (1) problem identification and formulation; (2) exploration, formalization, and problem solving; and (3) decision dissemination and implementation" (p. 108). The second phase has an impact on the number, nature, and effectiveness of solutions generated, as well as the scope and content of the discussions. If one focuses on differences of degree among group members in any factor, one will find that groups consisting of members at higher levels of "knowledge and ability" will out-perform those with lower levels when the shared task involves creative problem solving. When the factors studied are different in kind, when faced with "complex, non-routine problems," the more "variety in skills, knowledge, abilities and perspective," the more effective the team (p. 109).

Regarding team composition and its effect on the exchange of information, diverse groups in terms of organizational experience often suffer from limited communication. While there is an assumption that age and organizational tenure are positively correlated, demographic evidence dispels it. The authors caution researchers "to separate the effects of age and tenure because explanations differ for why age and tenure might be related to innovation" (p. 110). Their study resulted in the following demographic factors of team composition affecting a team's ability to be innovative: age, organizational tenure, educational background, and functional experience, but ultimately, their "results suggest support for the cognitive resources perspective, which posits that both resource level and diversity are important for innovation" (p. 120).

The Role of the Champion

Howell and Higgins (1990) take a different perspective on demographics, seeking empirical evidence to support the hypothesis of champions, individuals within an organization who take risks by introducing new ideas and innovative techniques to a group, process, or industry to promote their ideas. They sought evidence that "personality characteristics, leadership behaviors, and influence tactics...influence the emergence of champions in organizations" (p. 318). Their study relied on previous literature in entrepreneurship, transformational leadership, and influence, as there is a perceived positive correlation between entrepreneurs and champions. Champions often inspire others, and therefore they are perceived as leaders even though they do not have the hierarchical stature or title.

Howell and Higgins developed questionnaires and survey instruments that were completed by 25 matched pairs of perceived champions and non-champions, seeking responses concerning the personality characteristics, leadership behaviors, and influence tactics of champions of technological innovations. In sum, "champions exhibited higher risk taking and innovativeness, initiated more influence attempts, and used a greater variety of influence tactics than non-champions [and] showed that champions were significantly higher than non-champions on all paths in the model" (p. 317).

As the business environment becomes increasingly complex and competitive, finding leaders who can contribute positively to technological innovation—its identification, evaluation, and adoption—is an essential factor in productivity, competition, and survival. Success is in part reliant on there being a champion, someone who is not necessarily a corporate figurehead, but more often an individual who can contribute to an organization's success by promoting significant ideas and methods in an enthusiastic manner at critical points in that firm's or in a specific product's lifespan. As previous research has shown, "in order to overcome the indifference and resistance that major technological change provokes, a champion is required to identify the idea as his or her own, to promote the idea actively and vigorously through informal networks, and to risk his or her position and prestige to ensure the innovation's success" (p. 317). Champions should not be confused with gatekeepers, as their roles are quite different. The gatekeeper moves information from external sources to the cohorts within a project that will benefit by it. The gatekeeper is not necessarily a member of any of the project groups. Champions, on the other hand, as members of a project's cohorts, go outside the project and its sponsoring organization to find external information deemed beneficial, then uses their enthusiasm for the project as a tool aimed at convincing peers as to the use of the new idea.

Not all enthusiasts are champions. There are personality traits that contribute to one's ability to be a champion, including those resulting in being able to lead and influence others. Previous research has shown that "since innovation adoption is largely a process of influence...both with subordinates, as indicated by leadership behavior, and with peers and superiors, as indicated by influence tactics," champions are essential to innovation adoption (pp. 318-320). Transformational leaders, sometimes referred to as change agents, "use intellectual stimulation to enhance followers' capacities to think on their own, to develop new ideas, and to question the operating rules and systems that no longer serve the organization's purpose or mission" (p. 321). They boost the confidence levels and enhance the skill sets of their subordinates to the degree of their assisting in promoting and executing innovative strategies in response to challenges.

Howell and Higgins reviewed the literature on entrepreneurial personalities and found "that entrepreneurs desired to take personal responsibility for decisions, preferred decisions involving a moderate degree of risk, were interested in concrete knowledge of the results of decisions, and disliked routine work," and are often involved in innovative activities (p. 322). This is a profile consistent with that of innovators who display confidence in their abilities and seek opportunities to test and demonstrate them.

The recipe for a champion is a combination of personality traits such as enthusiasm and risk-taking, leadership skills strong enough to move people from accepted and comfortable practices to new and untested behaviors, and organization-wide vision that can result in the adoption of innovative practices. The authors believe that their study, as a concatenation of empirical evidence and established theory "in the domains of entrepreneurship, transformational leadership, and influence...provide organizational implications for the detection, selection, and development of champions." Champions are risk-takers, innovators, and informal transformational leaders, in their view, as opposed to what other researchers have suggested. It is not "that dispositional effects are less important than situational effects in influencing people's attitudes in organizational settings, [but that] our results suggest that by ignoring individual differences, one neglects major variables relevant to an important organizational human resource" (p. 336).

Functional Diversity

Anconia and Caldwell (1992) seek to understand the impact of functional diversity on product development teams, challenging the accepted wisdom that, as the complexity of tasks suggests an increase of the use of cross-functional and diverse teams, innovation in product development will be enhanced by combining engineers with other specialists. The authors looked at the performance of 45 high-tech new product development teams by integrating "group demography with other aspects of group theory to predict [their] performance." They looked at "the direct effect of group demography and the indirect effects created by demography's influence on internal processes and external communication" (p. 322). Specifically, their "study investigates two things: the direct effects of group heterogeneity on ratings of new product team performance, and the indirect effects of heterogeneity attributable to internal group process and to communication with non-group organization members" (pp. 325-326).

Their findings include different effects resulting from functional diversity (job tasks, domains, and limitations) and tenure diversity (length of time in a position) on team communication. The more functional diversity there was, the more team members communicated outside of their teams' boundaries, with groups such as marketing and management. Moreover, "the more the external communication, the higher the managerial ratings of innovation" (p. 321). When looking at tenure diversity, its effects were centered on how team members interacted rather than how and how much they communicated with external groups. The more diverse the tenure, the greater the level of improvement in group dynamics such as setting and buying into group goals and priorities, which yielded positive team assessments of performance. Sometimes, though, diversity is not wholly good and not solely positive, as it can also directly impede team performance. As the authors found, "Overall the effect of diversity on performance is negative, even though some aspects of group work are enhanced. It may be that for these teams diversity brings more creativity to problem solving and product development, but it impedes implementation because there is less capability for teamwork than there is for homogeneous teams" (p. 321). It is up to the team to navigate the negative functions of diverse teams so that the positive effects that diversity has on team interaction can be realized; this can be accomplished through increased communication and proactive resolution of conflict. Concurrently, it is up to management to recognize the tensions and impediments of diverse team composition so that the team does not suffer from organizational pressures from above, and the team is rewarded for its combined outcomes rather than individuals' functional outcomes.

Previous research has established that different demographic factors affect different variables. Age and sex affect economic states, geographic movement patterns, and crime rates, for example. How do demographic factors affect organizations and teams? Researchers have related demographic composition to employee turnover, to supervisors' performance evaluations of subordinates, and to types and levels of innovative practice within organizations. "For product development teams, two variables are likely to be of particular importance: the homogeneity of organization tenure and the mix of functional specialties" (p. 322). If a team is cross-functional, it has greater access to a wider array of expertise than if it were uni-functional. If a cross-functional team includes representatives from other departments of an organization, such as marketing, product transfer is facilitated. The authors also suggest that similarity in organizational tenure leads to enhancement of team integration because of increased "attractiveness of members to one another" (p. 323). Previous research also indicates that demographic diversity results in more conflict, and less team cohesion and ineffective communication than a homogenous team. Research on team conflict indicates that there will be more conflict among team members when individuals are relying on each other for task completion but they share different goals; the literature on group formation and effectiveness indicates that teams with diverse members often have difficulty integrating different values and cognitive styles. Their conclusion is that, if "not managed effectively, this diversity can create internal processes that slow decision making and keep members from concentrating on the task" (p. 323). Tenure diversity, for a new team member, also poses such negative factors as fewer opportunities to interact, differences in experience and perspectives, and a sense of exclusion from those group members who were teamed together at the same time. These distinctions hinder a team's ability to establish goals or set priorities.

The research literature on innovation also finds that diverse teams suffer from difficulties in reaching agreement or setting goals and establishing priorities. As new product teams have to obtain and share resources and information from other entities within an organization so that their tasks and goals can be accomplished, understanding the role of team communication—both internal and external—is important. High performing teams have greater amounts of communication among team members than low performing teams and organizations. If a team has a member that can translate and transmit information from the outside to the other team members, group performance can be enhanced.

The authors conclude by indicating that team productivity can increase when a team is composed of people with diverse sets of skills, knowledge, and interpersonal relationships with others in the organization. Tenure diversity can also be a positive factor, if the team takes advantage of its "range of experiences, information bases, biases, and contacts. Members who have entered the organization at different times know a different set of people and often have both different technical skills and different perspectives on the organization's history" (p. 325). However, "from a managerial perspective these research findings suggest that simply changing the structure of teams (i.e., combining representatives of diverse function and tenure) will not improve performance. The team must find a way to garner the positive process effects of diversity and to reduce the negative direct effects. At the team level, training and facilitation in negotiation and conflict resolution may be necessary. At the organization level, the team may need to be protected from external political pressures and rewarded for team, rather than functional, outcomes. Finally, diverse teams may need to be evaluated differently than homogeneous teams" (p. 338).

Individual Adaptation

Unlike Anconia and Caldwell, Scott and Bruce (1992) focus on the individual and his or her influence on and adaptation to an organization's climate for innovation. The authors ground their research in social interactionist theory and address the questions of how "leadership, work group relations, and problem-solving style affect individual innovative behavior directly and indirectly through perceptions of a "climate for innovation" (p. 581). Does the actual job or task that an individual is working on influence innovative behavior? Previous research has indicated that there is a "moderate relationship between climate and performance." To ascertain whether or not a job task proves to be a facilitator or hindrance to creativity, the authors "tested whether type of job assignment moderated the relationship between innovative behavior and each of the predictors in the model" (p. 581). They hypothesize that an individual's perception of the organizational climate for innovation is affected by the variables of leadership, work group relations, and problem-solving style, and used structural equation analysis to determine that the model proposed explained 37% of the variance in innovative behavior, concluding that task type actually does moderate the relationship between leader role expectation and innovative behavior (p. 580).

Referring to the accepted belief that managing innovation grows more difficult as an individual's attention is diverted from assigned task or goal, understanding both the motivation for innovative behavior and the nature of an environment that can foster motivation is essential. "Managing attention is difficult because individuals gradually adapt to their environments in such a way that their awareness of need deteriorates and their action thresholds reach a level at which only crisis can stimulate action" (p. 580).

There are three stages of individual innovation: problem recognition and the beginnings of solution generation, seeking of sponsorship or buy-in for that solution from those within the innovative environment, and the production of a model or prototype that can be used by a wider array of stakeholders. Innovation turns out to be the outcome of four interacting systems: an individual, a leader, a work group, and the climate for innovation (p. 582). What distinguishes Scott and Bruce's research from the literature addressing the relationship among tasks, technology, perception of climate, and effectiveness is their attention to how a task affects the moderation of the climate of innovation at the individual level. "When a task is routine or when individual discretion is low, the relationship between climate and innovative behavior is likely to be weaker than when the task is non-routine and high discretion is granted" (p. 588). Their summary of findings includes "leadership, support for innovation, managerial role expectations, career stage, and systematic problem-solving style to be significantly related to individual innovative behavior, and the hypothesized model explained almost 37 percent of the variance in innovative behavior" (p. 600).

The Innovative Process

Hage and Dewar (1973) elucidate the practicality of "elite values" as a concept and discuss how to go about measuring them empirically. "There are both formal and behavioral criteria, and one can easily imagine a continuum of elite participation with the precise cutting point of membership and non-membership remaining indistinct. Therefore, various definitions of membership need to be explored" (p. 279). To do so, the authors juxtapose for comparison concepts of elite values and those identified as criteria for leaders and members with the "three structural variables of complexity, centralization, and formalization" (p. 279). The amount of variance among these terms supported their being considered as independent variables. Hage and Dewar present the hypothesis that what the elites of an organization value has a direct effect upon what the organization accomplishes, but complicate the question by suggesting four alternative explanations for an organization displaying behaviors associated with elite values.

The first explanation involves organizational structure, as previous research has shown that structural variables such as complexity, centralization, and formalization are related both to rates of program change and the prediction of other organizational performances. One should also consider task structure, the second explanatory idea, as an organization maintaining a diversity of tasks will have a diversity of perspectives applied to them. This interplay of task and perspective "produces a creative dialectic that results in the development of innovative products and services," as a diversified task structure necessitates a diversified pool of professionals, each having access to specialized knowledge and sources of information (pp. 279-280). Third, Hage and Dewar note the "inverse relationship between centralization and program innovation." The more influential, elite members of an organization who can effect change and retain that power, the less opportunity there will be for non-elites to implement innovative ideas, as their suggestions would be concomitant with suggestions for change in "power, privilege and reward." Fourth, if power is centralized, there is less opportunity for innovation resulting from exchange of ideas or "the creative dialectic implied by complexity or diversity of tasks" (p. 280). Hage and Dewar gauged the degree of centralization by asking staff to describe the amount of participation they had in decision making, aggregating their responses as an average of positional means, with staff classified according to task assignment and hierarchical position in the organization (p. 283), suggesting that there is complementarity between elite size and degree of centralization.

There is a truism in organizational research that leaders are the source of strong influence. The authors describe leadership in organizations as "an interactive process wherein the leader provides certain services to the organization in exchange for legitimacy, respect, and compliance with his wishes by the staff" (p. 280). But leaders also act as mediators, spokesmen, and decision makers, and therefore the value they bring to an organization can be summarized as an agent of uncertainty reduction; power and influence increase as the general level of uncertainty among the staff decreases.

The authors conclude that ultimately, the values held by the elite group are more significant when predicting innovation than the values of any single leader or even the entire staff, but the correlation between a single leader and innovation should not be dismissed as a valid predictor of an organization's ability to innovate. "The association between Elite values and performance gives one some basis for arguing that statements about the goals of an organization from members of the Elite are probably more adequate than the executive director's perceptions alone" (p. 287).

A Dual-Cored Model of Organizational Innovation

Daft (1978) investigated the innovative process in an educational setting and examined the "behavior of administrators' vis-à-vis lower employees as innovation initiators...for a sample of school organizations [and relates his findings] to the professionalism of organization members, organization size, and frequency of innovation adoption" (p. 194). What leads administrators and other technical employees in this domain to adopt innovative approaches to problem solving? Daft offers findings of previous research to support the theory that there can be opposing innovative processes in an organization: one that begins at the lower levels of its hierarchy, and one that percolates down from upper levels. Therefore, his "findings are used to propose a dual-core model of organizational innovation" (p. 193).

Do organization leaders have primary impact on organizational innovation, as Hage and Dewar (1973) propose, linking innovation adoption to the "status and sociometric centrality of organization top administrators," or to top administrators' "cosmopolitan orientation," or perhaps with individual administrator's "motivation to innovate" (Daft, 1978, p. 193)? The answer is elusive, but much of the accepted research on innovation to the time of Daft's article suggests that leaders are active in the innovation process, acting as the connection between the organization's hierarchy and the technical environment in which staff and administrators work. By mere status and rank, organizational leaders are in position to innovate, and they can serve in supporting roles as well, by finding funds to implement new programs. By virtue of their leadership status, top administrators also influence innovation by setting priorities and goals (p. 193). Daft's concern is to find "underlying organization processes" that support innovation, as such information would be beneficial in providing knowledge that supports innovative alternatives to existing problems, and basic knowledge of organizational innovative processes would facilitate new ideas that may eventually provide innovative techniques that can be adopted (p. 194).

The two components of Daft's dual-core model derive from the organizational/human and the technological/mechanical orientations of a firm. This dichotomy is illustrative of how innovative ideas can come from either end of an organization's hierarchy, with administrative innovations moving from top to bottom, and technical innovations moving from bottom to top; therefore, innovative "ideas follow different paths from conception to approval and implementation" (p. 195). Keeping in mind that the larger an organization is, the more complex and refined the division of labor, it is not unusual that technical staff will be most concerned with technical innovations that are within their scope of work. This is in keeping with the elements of professionalism that define this group as a working unit, as they base the esteem they offer on team members' "education and training, participation in professional activities, exposure to new ideas, autonomy, and the desire for recognition from peers rather than from the formal hierarchy" (p. 196).

Daft summarizes the different roles that administrators and technical staff play in the innovation process, with the basic distinctions in definition of the groups being that a "technical innovation is an idea for a new product, process or service. An administrative innovation pertains to the policies of recruitment, allocation of resources, and the structuring of tasks, authority and reward. Technical innovations usually will be related to technology, and administrative innovations will be related to the social structure of the organization." This being the case: "(1) Each group is expected to initiate innovations pertaining to their own organization task; (2) this division of labor is expected to heighten as employee professionalism and organization size increase; (3) the absolute number of proposals initiated by each group is also expected to increase as professionalism and size increase; but (4) the greater number of proposals may not lead to greater adoptions because professionalism and size may be associated with greater resistance to adoption" (p. 197).

This does not mean that there are barriers between the two domains in terms of initiation and development of innovative strategies and solutions. In both the administrative and technical components of an organization, the process of innovation is often based on the professionalism of the employees and offered by individuals who are domain experts who will most likely use and benefit from the innovation proposed. In some organizations, "the amount of innovation and the degree of coupling between the two cores may be a function of technology, rate of change, and uncertainty in the environmental domain as well as employee professionalism" (p. 206). However, Daft finds that administrative innovation tends to happen in anticipation of changes in factors in the administrative domain, such as new goals and objectives, hierarchy, and control structures. In sum, organizations usually adopt more administrative than technical innovations, and the "technical core appears to be subordinate and tightly coupled to an active and influential administrative core" (p. 207).

Innovation Success Factors

Quinn's 1985 article opens a window onto the results of a multi-year, worldwide study of innovative companies, and highlights some of the "similarities between innovative small and large organizations and among innovative organizations in different countries" in an effort to understand the pervasive perception that world technological leadership is passing from the United States to our international rivals in Europe and the Far East (p. 73). While some would hold bloated U.S. corporate bureaucracies responsible for stifling innovation, Quinn suggests that there are large companies that understand what it takes to promote the innovative process and reap its rewards, like so many entrepreneurs who accept "the essential chaos of development, pay close attention to their users' needs and desires, avoid detailed early technical or marketing plans, and allow entrepreneurial teams to pursue competing alternatives within a clearly conceived framework of goals and limits" (p. 73). Small companies are associated with successful production of innovative ideas and products for several reasons, including that "innovation occurs in a probabilistic setting." Quinn refers to the very high percentages of venture failure—failures that the general public and corporate competitors never see—as one reason. On the opposite end of the spectrum are large corporations that might want to promote a new concept or product but are limited by the knowledge that innovation brings with it the real costs of failure. Unlike a new, small business, a large corporation does not want to "risk losing an existing investment base or cannibalizing customer franchises built at great expense" or attempt to "change an internal culture that has successfully supported doing things another way" or dismiss "developed intellectual depth and belief in the technologies that led to past successes." Emergent companies are not scrutinized and restricted by "groups like labor unions, consumer advocates," and should their venture fail, they "do not face the psychological pain and the economic costs of laying off employees, shutting down plants and even communities, and displacing supplier relationships built with years of mutual commitment and effort" (pp. 73-74).

Quinn's findings, supported by previous research in the management of innovation, yield a list of "factors [that] are crucial to the success of innovative small companies," including:

1. **Need orientation**: A personality trait held by inventor entrepreneurs. "They believe that if they 'do the job better,' rewards will follow."
2. **Experts:** *and fanatics* are two adjectives applicable to company founders "when it comes to solving problems."
3. **Long time horizons:** Unlike their large corporation counterparts, inventor entrepreneurs, fanatics that they are, tend to "underestimate the obstacles and length of time to success. Time horizons for radical innovations make them essentially 'irrational' from a present value viewpoint."
4. **Low early costs:** Innovators often are not burdened by large fixed costs such as rent, payroll, and employee health care plans, tending to minimize expenses by maximizing existing space.

5. **Multiple approaches:** "Technology tends to advance through a series of random—often highly intuitive—insights frequently triggered by gratuitous interactions between the discoverer and the outside world. Only highly committed entrepreneurs can tolerate and even enjoy this chaos."
6. **Flexibility and quickness:** Two traits associated with and resultant from each of the items above, with the added benefit of being undeterred "by committees, board approvals, and other bureaucratic delays, [allowing] the inventor-entrepreneur [to] experiment, test, recycle, and try again with little time lost."
7. **Incentives:** While small-scale innovators realize that they will personally reap the benefit of their success, Quinn recognizes that they "often want to achieve a technical contribution, recognition, power, or sheer independence, as much as money."
8. **Availability of capital:** Quinn's assessment at the time of publication posits "America's great competitive advantages [to be] its rich variety of sources to finance small, low-probability ventures" (pp. 75-76).

Juxtapose these eight factors for innovative success against what Quinn terms "bureaucratic barriers to innovation," and one has a comprehensive picture as to how and why many in 1985 viewed America as losing its role as the world's technological leader:

1. **Top management isolation:** Similar to what Bantel and Jackson would posit five years later, Quinn suggests that "senior executives in big companies have little contact with conditions on the factory floor or with customers who might influence their thinking about technological innovation."
2. **Intolerance of fanatics:** As is often the case in many domains where an unknown challenges the status quo, Quinn finds that large "companies often view entrepreneurial fanatics as embarrassments or troublemakers."
3. **Short time horizons:** Ladened with the weight of public perception and stockholder expectation, corporations often feel the "need to report a continuous stream of quarterly profits," which is antithetical to "the long time spans that major innovations normally require."
4. **Accounting practices:** "By assessing all its direct, indirect, overhead, overtime, and service costs against a project, large corporations have much higher development expenses compared with entrepreneurs working in garages."
5. **Excessive rationalism:** Large entities need to be managed and often via cautious business practices that are prescriptive and constrained in comparison to free-wheeling entrepreneurial techniques. "Rather than managing the inevitable chaos of innovation productively, these managers soon drive out the very things that lead to innovation in order to prove their announced plans."

6. **Excessive bureaucracy:** As a direct result of excessive rationalism, "bureaucratic structures require many approvals and cause delays at every turn."
7. **Inappropriate incentives:** The process of innovation is rife with surprise, which is poisonous to the "reward and control systems in most big companies." Large, bureaucratic organizations cannot permit challenges to their "well-laid plans, accepted power patterns, and entrenched organizational behavior" (pp. 76-77).

Quinn proposes that corporate executives understand and adapt to the fact that the innovation environment is filled with surprise, characterized by chaos and virtually immune to control. If they begin to adopt and implement some of the techniques and perspectives that successful entrepreneurs practice, and bring "top-level understanding, vision, a commitment to customers and solutions, a genuine portfolio strategy, a flexible entrepreneurial atmosphere, and proper incentives for innovative champions, many more large companies can innovate to meet the severe demands of global competition" (p. 84).

IT Use for Competitive Advantage

Porter and Millar (1985) address the role of information technology in an organization's strategy for competitive advantage. It is not insignificant or obvious to point out that at the time of this article's publication, information technologies fulfilled primarily quantitative functions. 1985 preceded ubiquitous e-mail, graphical user interfaces, and the Internet as a global communication network. As the authors indicate, "Until recently, most managers treated information technology as a support service and delegated it to EDP departments." Porter and Millar's goal is to "help general managers respond to the challenges of the information revolution" such as it was in 1985. Presciently enough, they are quick to point out that "managers must first understand that information technology is more than just computers." Indeed, understanding how managers can put these tools to use for competitive advantage is essential, as IT is "transforming the nature of products, processes, companies, industries, and even competition itself" (p. 149).

How should managers view information technologies in order to assess and implement their strategic significance? How and why do these technologies change "the way companies operate internally as well as altering the relationships among companies and their suppliers, customers, and rivals"? They should begin by recognizing the general ways in which technology affect business competition: "it alters industry structures, it supports cost and differentiation strategies, and it spawns entirely new businesses." Once these outcomes have been recognized, managers can follow the authors' five-step plan to "assess the impact of the information revolution on their own companies." Looking back with over 20 years of growth and change in information technologies, it is to the authors' credit for indicating early on how business advantage will derive from our abilities to make strategic use of the increasingly convergent and linked technologies that process the information (p. 149).

Porter and Millar urge managers to understand that the "role of information technology in competition is the 'value chain,' [a] concept [that] divides a company's activities into the technologically and economically distinct activities ['value activities'] it performs to do business." Profitability derives from a business's ability to create more value than "the cost of performing the value activities." Competitive advantage is achieved when "a company...either perform[s] these activities at a lower cost or...in a way that leads to differentiation and a premium price (more value)" (p. 150).

One can assess the potential profitability of a company by understanding its structure, which Porter and Millar contend is "embodied in five competitive forces...the power of buyers, the power of suppliers, the threat of new entrants, the threat of substitute products, [and] the rivalry among existing competitors." The degree to which the strength of these five forces coalesces obviously "varies from industry to industry as does average profitability." What managers need to keep in mind, however, is that the "strength of each of the five forces can also change, either improving or eroding the attractiveness of an industry" (p. 155). Moreover, given that "information technology has a powerful effect on competitive advantage in either cost or differentiation...technology affects value activities themselves or allows companies to gain competitive advantage by exploiting changes in competitive scope" (p. 156).

What specific steps can managers take to avail themselves and their companies of the strategic advantages for which information technologies have already provided positioning? Porter and Millar offer the following five:

1. **Assess information intensity:** "A company's first task is to evaluate the existing and potential information intensity of the products and processes of its business units."
2. **Determine the role of information technology in industry structure:** "Managers should predict the likely impact of information technology on their industry's structure."
3. **Identify and rank the ways in which information technology might create competitive advantage:** "The starting assumption must be that the technology is likely to affect every activity in the value chain."
4. **Investigate how information technology might spawn new businesses:** "Managers should consider opportunities to create new businesses from existing ones."

5. **Develop a plan for taking advantage of information technology:** "The first four steps should lead to an action plan to capitalize on the information revolution" (pp. 158-159).

Ultimately, "companies that anticipate the power of information technology will be in control of events. Companies that do not respond will be forced to accept changes that others initiate and will find themselves at a competitive disadvantage" (p. 160).

CONCLUSION

A laboratory that places a reliance on internal information, while excluding outside information from flowing into the lab, will ultimately fall short of its potential for discovery. This is demonstrated by the success of researchers inside a laboratory, who make greater use of individuals outside the organization and/or the literature (Allen & Cohen, 1969). This work lays a critical foundation for the need to boundary-span and the existence and importance of different types of roles within a research environment, and for organizations involved with innovation in general. These boundary roles satisfy an organization's communication network's role of bridging an internal information network to external sources of information (Tushman, 1977). Boundary roles and the interface between employees inside an organization with employees outside of an organization are highlighted in the discussion of the importance of informal trading of proprietary know-how between members of rival (and non-rival) firms (von Hippel, 1987). This work arguably sets the theoretical ground for the open source movement, while Freeman's (1991) consideration of benefit derived from industry-wide cooperation and networks of innovators similarly sets the theoretical stage for industry/government consortia focused on pre-competitive research. While the importance of boundary spanning and working/interacting with individuals outside the organization has been discussed, the criticality of close proximity is stressed, due to the "stickiness of information" (von Hippel, 1994).

Enormous structural changes in an organization are necessary to initiate and promote innovative thinking (Thompson, 1965). The focus on structural change is to move an organization away from one being highly efficient but low on innovation, to one that is high on innovation while retaining as much efficiency as possible. Some of these changes are illustrated by findings associated to the nature of team structures that support innovation. Leadership characteristics that result in a tendency to innovate are suggested to be a function of the top management team, rather than the CEO (Bantel & Jackson, 1989). A critical part of the management team, in relation to a specific innovation, is that of the *champion*. Champions are identified as the key individuals within an organization who take risks by introducing new ideas and innovative techniques to a group, process, or industry to promote their ideas (Howell & Higgins, 1990). Moving from managers to team members, it was found that the more functional diversity in an organization's product development team, the more team members communicated outside of their teams' boundaries with groups such as marketing and management. External communication corresponded to higher managerial ratings of innovation (Anconia & Caldwell, 1992). Also important is focusing on the individual's influence on and adaptation to an organization's climate for innovation. It is found that organizational climate for innovation is affected by the variables of leadership, work group relations, and problem-solving style, concluding that task type moderates the relationship between leader role expectation and innovative behavior (Scott & Bruce, 1994). Having considered teams, some final thoughts are offered on the innovative process.

There is complementarity between elite size and degree of centralization, and this is important for the acceptance and integration of innovation (Hage & Dewar, 1973). In other words, the upper management or elite are the enablers or disablers of innovation. This view is expanded on from findings that there can be opposing innovative processes in an organization: ones that begin at the lower levels of an organization's hierarchy, and ones that percolate down from upper levels—a dual-core model of innovation (Daft, 1978). The consideration of the differences associated with firm, size, region, and incumbent vs. emergent companies is offered though a multi-year study (Quinn, 1985). Finally, early advice was offered on the role of information technology in obtaining competitive advantage (Porter & Millar, 1985):

1. Evaluate the existing and potential information intensity of products and processes.
2. Predict the likely impact of information technology on industry structure.
3. Identify and rank the ways that information technology may offer competitive advantage.
4. Embrace the idea that information technology promotes the spawning of new businesses from existing ones.

REFERENCES

Allen, T.J., & Cohen, S.I. (1969). Information flow in research and development laboratories. *Administrative Science Quarterly, 14*(1), 12-19.

Anconia, D.G., & Caldwell, D.F. (1992). Demography and design: Predictors of new product team performance. *Organization Science, 3*(3), 321-341.

Bantel, K.A., & Jackson, S.E. (1989). Top management and innovations in banking: Does the composition of the top team make a difference. *Strategic Management Journal, 10*(2), 107-124.

Daft, R.L. (1978). A dual-core model of organizational innovation. *Academy of Management Journal, 21*(2), 193-210.

Freeman, C. (1991). Networks of innovators: A synthesis of research issues. *Research Policy, 20*(5), 499-514.

Hage, J., & Dewar, R. (1973). Elite values versus organizational structure in predicting innovation. *Administrative Science Quarterly, 18*(3), 279-290.

Howell, J.M., & Higgins, C.A. (1990). Champions of technological innovation. *Administrative Science Quarterly, 35*(2), 317-341.

Myers, S., & Marquis, D. (1969). *Successful industrial innovation.* Washington, DC: National Science Foundation.

Quinn, J.B. (1985). Managing innovation: Controlled chaos. *Harvard Business Review, 63*(3), 73-84.

Porter, M.E., & Millar, V.E. (1985). How information gives you competitive advantage. *Harvard Business Review, 63*(4), 149-160.

Scott, S.G., & Bruce, R.A. (1994). Determinants of innovative behavior: A path model of individual innovation in the workplace. *Academy of Management Journal, 37*(3), 580-607.

Thompson, V.A. (1965). Bureaucracy and innovation. *Administrative Science Quarterly, 10*(1), 1-20.

Tushman, M.L. (1977). Special boundary roles in the innovation process. *Administrative Science Quarterly, 22*(4), 587-605.

von Hippel, E. (1987). Cooperation between rivals: Informal know-how trading. *Research Policy, 16*(1), 291-302.

von Hippel, E. (1994). "Sticky information" and the locus of problem solving: Implications for innovation. *Management Science, 40*(4), 429-439.

Section II
Innovation, Influence, and Diffusion: Executive Summary

The chapters in this section introduce the seminal literature on technological diffusion, product diversification, and organizational strategies and constraints that firms encounter when introducing and adopting new technologies and innovative management strategies. The chapters also address the role of knowledge in the operation of organizations—both the effects of knowledge on organizational change and the effects of knowledge transfer within and among firms conducting innovative product design and development. Themes that run through this section include the rate and nature of change; attitudes, behaviors, and strategic change; and the role of research in organizational strategy. The closing chapter of this section addresses two kinds of new product development: the internal processes that assist or hinder development, and those that focus on factors that contribute to a new product's success or failure in terms of performance and diffusion.

The discussion of diffusion centers on distinctions between the processes undertaken by rational adopters of inefficient technologies and the conditions that promote the irrational rejection of efficient innovations. We also look at the use of patent and citation data as a method of gauging a firm's technological strength. In this case, information is the innovative product being diffused. Also addressed are diversification and organizational structure in terms of the identification and validation of factors that influence innovation adoption strategies. Structuration is introduced, focusing on the theory's social and historical substrata to provide an explanation of how we might rethink the roles of technology in organizations, and an overview of the literature on industrial innovation process is provided.

In Chapter VI, the role of knowledge in the operation of organizations—both the effects of knowledge on organizational change, and the manner and effects of knowledge transfer within and among firms conducting innovative product design and development—is addressed. The importance and processes of knowledge coordination within a firm's administrative hierarchy is brought to light, particularly through the concept of ba, or a shared space for the creation and emergence of knowledge. The role of radical change on cognition and its relationships to operative procedures and behavior norms are considered, as are methods to identify and expose organizational capabilities in the face of organizational structures that stifle innovation rather than institute and nurture change.

Knowledge transfer is a large subject addressed in this section, and we offer investigations into the reasons for and processes by which competing firms exchange organizational knowledge. Authors consider citation patterns of patent portfolios as a basis for comparing allied firms' technological strengths and assets to their output capabilities; they focus on the bio-technology industry to uncover how complex fields deploy experts; they give credence to the positive value that a resource dependency framework for innovation has for understanding and promoting effective interactions among constituencies when developing new products; and they look at the role of knowledge transfer and the product creation process to explain how concepts of modularity affect both product design and organizational design. The symbiotic relationship between technological innovation and its adaptation into the organizational environment is discussed, as is how research is organized in science and technology sectors. Chapter VII concludes with a view toward knowledge as an instrument of organizational change.

Chapter VII begins with a discussion of organizational innovation strategies: the interaction between technology and organizational structure, and how this kind of interaction affects how organizations function. Distinctions between radical and incremental outcomes fuel one innovation process model, suggesting that any organizational innovation process requires a unique implementation strategy and organizational structure that is responsive to the organizational conditions. We learn that investing resources into people as opposed to infrastructure can be a strong facilitator of innovation adoption, and that research from domains such as design, software engineering, and operations management can help us to understand how product architecture affects the performance of manufacturing firms.

How the values of upper management—and how an organization's members respond to change—affect the organization's use of process innovations is also brought forth, as are organizational constructs that relate determinants such as specialization, professionalism, and managerial attitudes toward change to innovation. Researchers provide analysis of industrial innovation in the United States, identifying the conditions that contribute to the success of new products, while others examine the methods, techniques, and tools (MTTs) that provide IS professionals with a quantitative basis for business process reengineering (BPR). Research's role in strategic change is the subject of the final section of the chapter, beginning with data gathered from 10 pharmaceutical companies that help in determining the role of 'competence' in that industry's research. Technological trends and other factors contributing to the nature of competition in organizations undergoing strategic change, factors such as the rate of technological change and diffusion, and the intensity of the role and importance of knowledge in an increasingly information-based global environment are addressed.

The articles discussed in Chapter VIII address the internal processes that assist or hinder development and those that focus on factors that contribute to a new product's success or failure in terms of performance and diffusion. Researchers examine the nature of the steps that affect the development process and determine how to improve process performance; the extent to which in-house parts development affects new product development and overall project performance; the tensions and trade-offs that occur among different functional areas and how they affect innovative product development; strategies to plan, focus, and control a firm's project development; and the many measures of product development success and failure over and against existing measures used by academic researchers. The section concludes with additional discussion of recent diffusion models of new product acceptance, helpful to both marketing managers and researchers; suggestions regarding how common and why it is that innovators find themselves in competition with product imitators who benefit more greatly than themselves; how we better understand the extent to which different industries make use of the patent systems to promote and protect innovation; and how product superiority is the number one factor influencing commercial success. Researchers provide insights on new product diffusion models in marketing, people's communicative behaviors, and innovation diffusion forecasting models and new product performance.

Chapter V
Diffusion and Innovation: An Organizational Perspective

INTRODUCTION

This chapter introduces the seminal literature addressing technological diffusion, innovative product diversification, and the organizational strategies and constraints that firms face when introducing and adopting new technologies and innovative management strategies. We begin with diffusion. Abrahamson (1991) draws critical distinctions between the processes undertaken by rational adopters of inefficient technologies and the conditions that promote the irrational rejection of efficient innovations. Attewell's (1992) focus is on organizational learning and abilities that drive the diffusion of innovative information and computing technologies. Cooper and Zmud (1990) examine managerial involvement with information technology, its effect on the adoption and infusion of that technology, and the role of rational decision models in explaining IT adoption. The section on diffusion closes with Narin and Perry's (1987) look at the use of patent and citation data as a method of gauging a firm's technological strength. In this case, information is the innovative product being diffused.

Chatterjee and Wernerfelt (1991) begin the second section on diversification and organizational structure by locating a theoretical basis for the identification and validation of factors that influence diversification innovation adoption strategies. Miller and Freisen (1982) ask why different decision-making variables affect entrepreneurial and conservative firms differently, focusing on the determinants of innovation that must be considered in any organization's development strategy. Orlikowski (1992) re-examines structuration to provide an alternative conceptualization of the role of technology, focusing on the theory's social and historical substrata to provide an explanation of how we might rethink the roles of technology in organizations. Rothwell (1992) concludes the second section by providing a rich overview of the literature on industrial innovation process, from which he soberly determines that, even after five decades of research on innovation in organizations, there is still no roadmap to successful innovation.

DIFFUSION

Rational Adopters and Inefficient Innovations

Whether we view the literature of innovation diffusion as self-fulfilling prophesy, accurate hind-sight, or the distillation of empirical evidence, Abrahamson, in his 1991 *Academy of Management Review* article, is intent on deflating the rosy picture that characterizes much of its content by providing a skeptic's perception of its fundamental premises. His aims are to identify the motivations, frequency, and nature of the processes that promote "technically inefficient innovations [to be] diffused or efficient innovations rejected" and to challenge the assumption that these ends are the result of "rational adopters [making] independent and technically efficient choices" (p. 587). Abrahamson's review of the literature complicates these seemingly straightforward ends he seeks by indicating "that processes, which prompt the adoption of efficient innovations, may coexist with processes that prompt the adoption of inefficient ones. Additionally, these resolutions inform research on the diffusion and rejection of many different types of innovations across varying contexts" (p. 586).

Guided by the dean of innovation diffusion literature, Everett Rogers, Abrahamson recounts Rogers' basic questions: "First, what processes and contextual factors affect innovations' rates of diffusion?" Do "theoretically derived mathematical models adequately describe longitudinal changes in diffusion rates"? "What characteristics differentiate earlier from later adopters?" Can one discern "differences between leaders and laggards" by analyzing the timing of technology adoption? "How does the structure of networks of adopters affect the sequence in which adoptions occur during diffusions?" (p. 586). Abrahamson cautions readers against blindly accepting what Rogers calls the "efficient-choice" perspective, "which assumes that rational adopters make independent and technically efficient choices, [and] dominates the innovation-diffusion literature," because it "perpetuates pro-innovation biases [by providing] limited help in addressing the questions of when and by what processes technically inefficient innovations are diffused or efficient innovations rejected"; often, "fads or fashions facilitate the adoption of technologies that are technically efficient for certain organizations, but not for many of those that adopt them" (p. 588).

As an example of this, Abrahamson cites Rumelt's 1974 study, finding "that diversification did not correlate with the adoption of a multidivisional structure after the early 1960s and concluded that structure follows not only strategy, but also fashion." Fads and fashions can also do damage to an organization's economic robustness if they "prompt rejections of administrative technologies that had the potential to become technically efficient for their adopters." Managers should be aware "that technologies become effective only through gradual, careful, and sustained implementation processes that provide organizations with tacit knowledge and the skills necessary to implement these technologies efficiently," and wary of "fads or fashions [that] cause organizations to leap rapidly from one technology to the next, so that no technology has enough time to work." In seeming opposition to "reviewers of the innovation literature [who] unanimously agree that it contains pro-innovation biases, such as Kimberly (1981), who defined pro-innovation biases as presumptions that innovations will benefit organizations," Abrahamson gets to the heart of the problem borne of his skepticism regarding fads and fashions as determinants of technology adoption. "Pro-innovation biases influence not only the questions that the innovation-diffusion literature emphasizes, such as what determines diffusion rates, but also the questions that the literature deemphasizes. These biases, for instance, suggest an obvious answer to the question: Why do innovations diffuse or disappear? Innovations diffuse when they benefit organizations adopting them, and they disappear when they do not" (Abrahamson, 1991, p. 589).

To correct the blanket assumptions to which researchers in this domain ascribe, Abrahamson points out that it "makes little sense, therefore, to ask what processes impel or counter the diffusion of innovations, when do these processes take hold, and to what extent do these processes cause the diffusion or rejection of innovations. It makes even less sense to ask whether certain processes diffuse non-beneficial innovations or cause the rejection of beneficial ones. Recognizing how pro-innovation biases limit questions addressed in the diffusion of innovation literature is important, but it does not explain how to research questions that do not reflect these biases. To do so, theorists must take three steps: First, they must examine how the dominant theoretic perspective in the innovation-diffusion literature contains assumptions that reinforce pro-innovation biases. Second, theorists must reject these assumptions in order to reveal counter-assumptions, which underlie less dominant perspectives that do not reinforce pro-innovation biases. Third, theorists must develop these less dominant perspectives in order to investigate questions that do not reflect pro-innovation biases" (pp. 589-590). Researchers in this field should concentrate more attention on "theories that maintain that life cycles deterministically impel innovations' progressions through preordained stages of invention, innovation, diffusion, maturity, and rejection, [which] have received little empirical support" (p. 593).

Technology and Organizational Learning

Unlike Abrahamson's focus on the roles and value of fads and fashions for innovative technology diffusion, Attewell's 1992 foray into the burgeoning field of business computing as a case study of the motivators and constraints of technology diffusion is far less significant than the analytical frameworks and analyses he employs to "construct a perspective on technology diffusion that places at its core the issue of organizational learning and know-how." The information that is brought to bear in support of his conclusions derives from how "survey, interview, and archival data on the recent diffusion of business computing are...analyzed, in order to demonstrate the empirical validity of [his] new theoretical formulation, and its utility in explaining institutional patterns of diffusion" (p. 1).

In opposition to what Attewell describes as the "dominant explanation for the spread of technological innovations, [which] emphasizes processes of influence and information flow," the author proposes an "alternative model which emphasizes the role of know-how and organizational learning as potential barriers to adoption of innovations." In the majority view, the more closely an organization's connection is to an innovative technology's pre-existing users, the more quickly the firm's adoption of that innovation. Moreover, the further away from the nexus of information flow a potential adopter is, the more slowly the adoption will be. In most cases, "Firms delay in-house adoption of complex technology until they obtain sufficient technical know-how to implement and operate it successfully. In response to knowledge barriers, new institutions come into existence which progressively lower those barriers, and make it easier for firms to adopt and use the technology without extensive in-house expertise." There is a positive correlation between the degree to which "knowledge barriers are lowered" and the rapidity with which "diffusion speeds up." For Attewell, therefore, "the diffusion of technology is re-conceptualized in terms of organizational learning, skill development, and knowledge barriers" (p. 1).

Attewell begins his discussion with Rogers' insight that "diffusion is a process of communication and influence whereby potential users become informed about the availability of new technology and are persuaded to adopt, through communication with prior users," establishing "patterns of adoption across populations of organizations [that] reflect patterns of communication flow." Attewell finds a second vehicle of comparison that researchers employ in the literature on diffusion, this time "an economic one which views diffusion primarily in terms of cost and benefit: the higher the cost, the slower diffusion will occur. The higher the perceived profit from an innovation, the faster adoption will occur" (p. 2). Combined, both comparative vehicles "inform one style of diffusion research which focuses upon adoption by individuals or by single firms." The characteristics of early adopters concern firm size, profitability, the presence or absence of "innovation champions," and other organizational and environmental attributes. In brief, "large firms adopt innovations before smaller ones...those firms for whom an innovation is most profitable become early adopters. [Champions play] three roles—the product champion, the business innovator, and the technological gatekeeper[, and the] intensity of competition, firm size, mass versus batch production, degree of centralization, organizational slack, proportion of specialists, and functional differentiation" are indicators of innovation adoption (p. 2).

Attewell, acknowledging Eveland and Tornatzky's (1990) "five elements of context" that promote and constrain choices related to adoption (the intrinsic nature of technology, user characteristics, innovation deployers' characteristics, the boundaries between innovative technology users and deployers, and communication and transaction mechanism characteristics), focuses on their "observations that diffusing or deploying a technology is more difficult if: (1) its scientific base is abstract or complex; (2) the technology is fragile; (3) it requires 'hand-holding' and advice to adopters after initial sale; (4) is 'lumpy' (affects huge swaths of the user organization); and (5) it is not easily 'productized,' made into a standard commodity or a complete package" (Attewell, 1992, p. 4).

Moving away from the classic research on diffusion theory, which was principally concerned with the amount and nature of contact between technology developers and technology users, and noting that "non-adopters lag behind early adopters because the former have not yet learned of the existence of an innovation, or have not yet been influenced about its desirability by better-informed contacts," Attewell pays careful attention to "the role of information and knowledge." Citing "Tushman and Anderson (1986), who "suggest that innovative technologies can either be competence-destroying or competence-enhancing for firms, according to whether they render obsolete or build upon preexisting skills and knowledge," he suggests that knowledge and learning, both on the individual and organization levels, are essential factors determining the success of technology adoption. "Neither 'learning by doing' nor 'learning by using' is the result of knowledge transfer from the originator to the user of the technology. Indeed the point of the concepts is the opposite: to highlight the need for learning and skill formation in situ, far from the originator" (Attewell, 1992, pp. 5, 6).

Traditionally, "the incentive to develop a new technology derives from the inventor's desire to monopolize the use of the innovation. The faster it diffuses, the sooner one's advantage and ability to profit from it go away." Attewell, attentive to the "economics of innovation...whether licensing arrangements, patents, joint ventures, and other special institutional arrangements intended to make it profitable for innovators to share their innovations" actually do so, makes clear the paradox technology innovators face in the competitive marketplace: "The existence of these special inducements to share knowledge underlines the fact that the initial inclination of businesses is to hoard and hide know-how, rather than transfer or diffuse it" (p. 6).

Management, IT, and Technology Diffusion

Unlike Abrahamson and Attewell's focus on who diffuses technology and why, Cooper and Zmud (1990) begin their *Management Science* article by providing a foundation for their examination of the "how": implementation research as it pertains to technology diffusion. Responding to "Kwon and Zmud's (1987)...IT implementation research model which is based on the organizational change, innovation, and technological diffusion literatures," Cooper and Zmud "apply this model in first framing and then interpreting the results of an empirical study examining the implementation of a common manufacturing IT application—that of material requirements planning" (p. 124). The authors are interested in "questions concerning the implementation of a production and inventory control information system (material requirements planning: MRP) [that] focus[es] on the interaction of managerial tasks with the information technology and the resulting effect on the adoption and infusion of that technology." Although Cooper and Zmud "find that this interaction does indeed affect the adoption of MRP...it does not seem to affect MRP infusion. These results support the notion that though rational decision models may be useful in explaining information technology adoption, political and learning models may be more useful when examining infusion" (p. 123).

Three perspectives are applied to the authors' examination of the firms in their study: "factors research, process research, and political research. Factors research focuses upon a variety of individual, organizational, and technological forces which are important to IT implementation effectiveness," with specific significant factors identified as "top management support of the implementation effort, good IT design, and appropriate user-designer interaction" (pp. 123-124). Process research involves "social change activities and suggests that implementation success occurs when commitment to change and the implementation effort exists, extensive project definition and planning occurs, and management of the process is guided by the organizational change theories." Finally, political research looks at how "the diverse vested interests of IT stakeholders affect implementation efforts, [noting] that successful implementation depends upon recognizing and managing this diversity" (p. 124).

Cooper and Zmud define information technology implementation "as an organizational effort directed toward diffusing appropriate information technology within a user community" (pp. 124-125). This effort takes the form of a process that includes several sequential steps: initiation, adoption, adaptation, acceptance, routinization, and infusion. For each of these steps, Cooper and Zmud describe both its process in terms of contributing to diffusion and the product of that process. Initiation involves reviewing an organization's challenges and opportunities in terms of the IT solutions that it has undertaken. The authors note that "pressure to change evolves from either organizational need (pull), technological innovation (push), or both." The result, or product, of this review process is the identification of a "match [that] is found between an IT solution and its application in the organization." Adoption's process is described as the "rational and political negotiations [which] ensue to get organizational backing for implementation of the IT application." The product is management's decision to "invest resources necessary to accommodate the implementation effort." Adaptation is the step in the diffusion process where the "IT application is developed, installed, and maintained, [demanding that organizational] procedures are revised and developed [and] members are trained both in the new procedures and in the IT application." The result of the adaptation process is an IT application [that] is available for use in the organization. Acceptance involves rallying organizational members to buy into using the IT application; acceptance is realized when the IT application is employed in organizational work. Routinization is the process of normalizing the IT application into the organization's regular activities, which produces an adjustment in the organization's governance systems...to account for the IT application." Ultimately, infusion takes place, in which "increased organizational effectiveness is obtained by using the IT application in a more comprehensive and integrated manner to support higher level aspects of organizational work." This yields the final product: "The IT application is used within the organization to its fullest potential" (pp. 124-125).

Cooper and Zmud refer to earlier research that "identified five major contextual factors which impact processes and products associated with each of these stages." These include characteristics of the user community such as "job tenure, education and resistance to change; characteristics of the organization, such as specialization, centralization, and formalization; characteristics of the technology being adopted, such as complexity; characteristics of the task to which that technology is being applied, including task uncertainty, autonomy and responsibility of [a] person performing the task, and task variety; and characteristics of the organizational environment, such as the level and nature of uncertainty and inter-organizational dependence" (pp. 124-125).

Acknowledging that their IT implementation model suggests that these characteristics are fluid and affect different stages of the implementation process to different degrees, Cooper and Zmud find that management must be aware "of the critical issues to be raised and resolved throughout the implementation process. It is only by adopting a comprehensive research framework and then examining sets of constructs from this framework in a systematic manner that substantial progress can be made in prescribing which issues should dominate for each of the IT implementation stages" (p. 136).

Innovation Diffusion through Research Citation

Narin and Perry (1987) take a step back from the operational contexts of Cooper and Zmud to present a hypothesis regarding the use of patent and patent citation data as a means of measuring a company's technological strength. They suggest that measuring these data against other performance indicators such as research and development budgets, peer evaluations, the number of research publications produced, and other quantifiable information such as increases in sales and profits, will prove to be an effective tactic in evaluating "the overall direction, breadth, and quality of a company's research program [and] potential for long-term corporate health" (p. 144).

To test their hypothesis, Narin and Perry examined quantitative data from 17 U.S. pharmaceutical firms to determine "which financial, R&D and expert opinion data were readily available." Their findings and its analysis indicate that "for these pharmaceutical companies...the patent data are an excellent indicator of overall corporate technological strength with (1) an overall correlation of 0.82 between expert opinion of pharmaceutical company technical strength, and the number of U.S. patents granted to the companies, and (2) correlations, in the general range of 0.6 to 0.9, between increases in company profits and sales, and both patent citation frequency and concentration of company patents within a few patent classes" (p. 143).

Emphasizing that the two different counts, patents and patent citation, "may reveal two different aspects of the research and development cycle, with patent counts indicating the size of research inputs and patent citation counts indicating the quality or impact of the research outputs," Narin and Perry come to the conclusion that "the use of patent citation data is one way to disentangle some of the company-to-company differences in patenting policies from the quality of company research programs" (pp. 154-155).

DIVERSIFICATION, INNOVATION, AND ORGANIZATIONAL STRUCTURE

Diversification

Chatterjee and Wernerfelt (1991) approach diversification with the intent to investigate how an organization uses its productive resources that it finds are supplemental to its current bread and butter operation. Their interest is in strategy, methodology, and implementation rather than uncovering why a particular firm puts its resources to particular goals. They posit that a firm understanding of these resources promotes management's ability to set strategies for future direction and expansion. They further claim that "excess physical resources, most knowledge-based resources, and external financial resources are associated with more related diversification, while internal financial resources are associated with more unrelated diversification." Chatterjee and Wernerfelt's goal is to establish a firm theoretical basis that will allow the identification of "systematic factors that influence the type of diversification, and empirically [examine] the validity of these factors to explain the type of diversification undertaken by a diverse group of firms between 1981 and 1985." A firm may diversify because it is "better at diversification strategy as well as industry selection, or it could be because some underlying factors allow them to enter these industries and make related diversification their best strategy" (p. 33). The firms' resource profiles, as they are portrayed at the beginning of the study, can be used to explain "the type of diversification" that their sample firms engaged in during the period of study. Moreover, the authors demonstrate that "high-performing firms conform more closely to the theoretical predictions than do low-performing firms" (p. 34). They seek "a quantifiable measure of the change in diversification profile for a sample of firms between 1981 and 1985" in order to validate their conclusions (p. 37).

Supporting their methodology is prior research suggesting that "a firm can gain...competitive advantages if it has skills or resources that it can transfer into the new market" and can benefit from diversification as "a means of managing resource-dependent relationships. The empirical evidence also suggests an association between diversification and the diversifying firm's resource position." Chatterjee and Wernerfelt begin by identifying: "(a) a typology of resources which is generalizable across different firms, and (b) the association between resources, type of markets, and the potential for value creation...[for this] resource-based approach allows us to adopt the perspective of the diversifying firm's managers." The more inflexible or product-specific a firm's resources, the greater constraint it experiences in terms of related diversification. Conversely, "if a firm possesses resources which are flexible (regarding end-products), it would have the option of either more or less related diversification." Chatterjee and Wernerfelt identify three classes of resources: physical, intangible, and financial. While the first two are relatively inflexible and are used exclusively for entry into related markets, financial resources, "being most flexible, are useful for any type of diversification" (p. 34).

We may consider physical resources such as manufacturing facilities and the equipment within it to limit an organization's fixed capacity of productivity. Any excess capacity would be directed to production in closely related industries. A firm's intangible assets include its recognition through brand names and logos, and even its ability to innovate, yielding what Chatterjee and Wernerfelt consider "'softer' capacity constraints." Because intangible assets are "also relatively inflexible, [they] can be used to most advantage in related industries." Financial resources, of which there are two classes, provide the greatest flexibility because "they can be used to buy all other types of productive resources...The first class, internal funds, consists of liquidity at hand and unused debt capacity to borrow at normal rates. The second class, external funds, consists of new equity and possibly high-risk debts (such as junk bonds)" (p. 35).

If "the profit potential of any firm depends on the resources it can control, looking at diversification as a way to leverage these resources [can indicate] how the type of diversification can lead to value creation." Chatterjee and Wernerfelt show a direct correlation between a firm's "pattern of diversification and the underlying resources of a firm." More often than not, "there is a strong association between intangible assets and more related diversification. There was no association between ability to raise equity capital and the type of entered market. We also found that higher performing firms supported the model better than lower-performing firms" (p. 46). Ultimately, success in terms of diversification rests not on applying "any one type of diversification...but it is the proper application of resources that will improve performance" (p. 36).

Conservative vs. Entrepreneurial Innovation Strategies

Miller and Freisen (1982) investigate why the "impact upon product innovation of environmental information processing, structural and decision making variables will vary significantly and systematically among entrepreneurial and conservative firms," and find that "future research on the determinants of innovation must consider organizational strategy" (p. 2). By examining empirical data on hand from 52 Canadian business firms, the paper seeks to present "distinct arguments concerning the determinants of innovation in conservative and entrepreneurial firms" through the analysis of "innovations in product lines, product designs, and services offered [and] not technological or administrative innovations." The authors suggest that their "arguments or findings may not hold for other types of organizations" (p. 2).

In brief, we should consider the conservative model of "product innovation as something that takes place only when absolutely necessary [and] assumes that innovation will not occur unless serious challenges in the environment are pointed out to and analyzed by managers who have resources sufficient for innovation." Conversely, the entrepreneurial model suggests that when "innovation resources are being wasted in the pursuit of excessive novelty...innovation will tend to be excessive and very high unless: (1) information processing systems alert executives to the dangers of too much innovation, and (2) analytical and strategic planning processes and structural integration devices do the same" (p. 1).

In an earlier study by Miller and Freisen (1980), the authors identify momentum as "a pervasive force in organizations; that past practices, trends and strategies tend to keep evolving in the same direction, [and] perhaps eventually reaching dysfunctional extremes...Centralization of authority often continues until the organization becomes an autocracy, while decentralization can lead to the proliferation of uncoordinated departmental fiefdoms." In that same study, Miller and Friesen "found that the same might be true of innovation. Firms with a propensity to innovate become still more innovative, sometimes passing the point of dramatically diminished returns. Conservative firms on the other hand sometimes drift towards complete stagnation" (Miller & Freisen, 1982, p. 2).

The preponderance of literature on product innovation is weighted toward "a conservative model of innovation [finding that innovation] is not a natural state of affairs, that it must be encouraged by challenges and threats, and that it requires a particular type of structure and an effective information processing system to make conservative managers aware of the need for change." Miller and Freisen "contend that the conservative model will apply to firms that perform very little innovation or risk taking" (p. 3).

The conservative model looks toward four kinds of prerequisites necessary to facilitate innovation. "First, there must be environmental challenges before innovation occurs. Second, there must be information about these challenges brought to key decision makers by effective scanning and control systems. Third, there must be an ability to innovate that is created by adequate resources, skilled technocrats, and structural devices. Fourth, there must be decision making methods appropriate for innovation projects" (p. 3).

Previous research indicates that "53 percent of the product and technological innovations...came in response to market, competitive, or other external environmental influences. The more dynamic and hostile (i.e., competitive) the environment, the greater the need for innovation and the more likely it is that firms will be innovative." Market-diverse firms "are likely to learn from their broad experience with competitors and customers. They will tend to borrow ideas from one market and apply them in another." There is a positive correlation between the quantity of product diversity in a given organization, "the probability that innovations will be proposed, and the...probability that organization members will conceive major innovations. Of course diversity in organization personnel, operating procedures, technologies and administrative practices increases with environmental heterogeneity." In addition, Miller and Freisen concur with earlier research in that "mechanistic structures impede innovation while organic structures facilitate it, in part because the former have much less information processing capacity" (p. 3). While an inability to scan the environment and "recognize the needs and demands of [a firm's] external environment" has a significantly negative effect on that organization's innovativeness, controls "that monitor task performance and financial results are said to identify areas of weakness and to prompt remedially oriented innovations" (p. 4).

Centralization, or the concentration of authority for decision making, is a structural variable associated with innovation. Significant concentration of power inhibits imaginative solutions to problems. The opposition of two powerful agents demands agile problem solving, and if power is concentrated it "often prevents imaginative solutions of problems." On the other hand, "dispersed power...can make resources more readily available to support innovative projects, because it makes possible a larger number and variety of sub-coalitions. It expands the number and kinds of possible supporters and sponsors" (p. 4). Differentiation and integration are also structural variables to consider in product innovation. "Unless there are integrative devices such as task forces, interdepartmental committees, integrative personnel, or matrix structures, collaboration is difficult and conflicts and mistakes result" (p. 5).

Decision-making variables involve the basis, manner, and degree to which executives in conservative firms "use and process information in innovative decision-making." These variables include the "degree of analysis of information, amount of planning, and the amount of explicit conceptualization of strategies" that are involved in innovation. "The more analysis is performed by key decision makers, that is, the greater the tendency to search deeper for the roots of problems and to generate the best possible solution alternatives, the more likely it is for innovation opportunities to be discovered and actualized." Additionally, Miller and Freisen refer to "futurity," a planning horizon that will "influence organizational innovation. The more future-oriented the firm, the greater the concern with change and therefore with innovation" (p. 5).

The last conservative firm variable that Miller and Freisen discuss is "the consciousness of strategy...the degree to which strategies have been explicitly considered and deliberately conceptualized." When an executive's focus is on non-strategic concerns, there will be a tendency to "muddle-through and [be] much less likely to engage in product innovation...but where there is a concerted attempt to decide upon the product-market orientation of the firm, there is a greater likelihood that target markets will be defined more broadly" (p. 5).

The alternative to the conservative model is the entrepreneurial, "which applies to firms that innovate boldly and regularly while taking considerable risks in their product-market strategies" (p. 5). Miller and Freisen make four points regarding the entrepreneurial model. First, there will be a high degree of innovation "unless good scanning or control systems reveal it to be too expensive or wasteful;" second, "effective analysis of decisions, futurity, and explicit and conscious considerations of strategy will also guard against the natural tendency towards innovative excess." Third, assuming that strategy is the strongest element driving innovation, "the role of environment as an innovation incentive will be reduced." And fourth, "the frequently observed positive covariance between innovation and structural factors such as technocratization and differentiation should prevail, but at a lower level of significance than for conservative firms" (p. 6).

Environmental variables, the degree to which any type of firm, entrepreneurial or conservative, is found to be in a dynamic or hostile environment, "are expected to relate positively to innovation. These environmental variables are particularly relevant to entrepreneurial firms "because their venturesome managers prefer rapidly growing and opportune settings; settings which may have high risks as well as high rewards." Heterogeneity is also positively correlated with innovation "because firms that innovate are more likely to come up with products and services that can be exploited in different markets." Innovation in entrepreneurial firms at times causes "dynamism, hostility, or heterogeneity, rather than the other way around." If and when that happens, "the greater latitude for strategic choice (e.g., to innovate in stable environments) will cause correlations between innovation and environment to be lower in entrepreneurial samples than in conservative samples" (p. 6).

As with conservative firms, "most structural variables are predicted to have a positive correlation with innovation in entrepreneurial firms." One caveat the authors suggest is "that, in general, the positive relationships between structure and innovation should be weaker in the entrepreneurial sample." Strategy and aggressiveness of management in entrepreneurial firms, rather than organizational structure as in conservative firms, contributes to "some entrepreneurial firms [having] a tendency to innovate a great deal even though their structures are less than ideal for this, according to the literature supporting the conservative model." The integrative variable is negatively correlated with innovation in entrepreneurial firms, the authors find. "Integrative devices such as committees, task forces, and integrative personnel bring important facts to bear upon decisions. The innovation proposals of enthusiastic but reckless executives are likely to be pared down by departments whose aim it is to ensure effective resource management and efficiency" (p. 7).

Miller and Freisen conclude their discussion of variables affecting product innovation by describing decision-making variables as they apply to entrepreneurial firms. Analysis, futurity, and consciousness of strategy "are expected to correlate negatively with the degree of product innovation...Analysis, planning, and the deliberate attempt explicitly to formulate strategies will provide the firm with a better knowledge of its opportunities and excesses. Any tendency to overindulge in product innovation may be curbed by these activities" (p. 7).

To summarize Miller and Friesen's findings: "The 'conservative' model views product innovation as something done in response to challenges, occurring only when very necessary. The model predicts that innovation will not take place unless:

1. There are serious challenges, threats, or instabilities in the environment;
2. These are brought to the attention of managers and consciously analyzed by them; and
3. Structural, technocratic, and financial resources are adequate for innovation.

In short, positive and significant correlations are expected of innovation with environmental, information processing, decision-making, and structural variables." When the 'entrepreneurial' model is considered, however, innovation is second nature, and a firm will "be boldly engaged in [it] unless there is clear evidence that resources are being squandered in the pursuit of superfluous novelty. The model postulates that innovation will tend to be excessive and extremely high unless:

1. Information processing (scanning and control) systems warn executives of the dangers of too much innovation, and
2. Analytical and strategic planning processes and structural integration devices do the same.

In other words, negative correlations of innovation with information processing, decision-making, and structural integration devices are expected" (pp. 16-17).

A New View of Structuration

Orlikowski (1992) moves past the opposition of conservative and entrepreneurial to build on Gidden's theory of structuration, and proposes that researchers concerned with the role of technology in organizational settings blend what was at the time of her writing a bifurcated and polarized perspective into a "structurational model," an approach that discourages "perspectives associated with technological research [that] have limited our understanding of how technology interacts with organizations...What is needed is a reconstruction of the concept of technology, which fundamentally re-examines our current notions of technology and its role in organizations" (p. 398). Orlikowski's "alternative theoretical conceptualization of technology...underscores its socio-historical context, and its dual nature as objective reality and as socially constructed product. This paper details and illustrates a structurational model of technology that can inform our understanding and future investigations of how technology interacts with organizations" (p. 423).

Standard approaches to research in this domain can be characterized as occurring in three periods. Orlikowski identifies the early research period as one which "assumed technology to be an objective, external force that would have deterministic impacts on organizational properties such as structure." She considers later research to shift perspective, concentrating on "the human aspect of technology, seeing it as the outcome of strategic choice and social action." Orlikowski's basic position is that neither of these perspectives can be considered wholly reliable, and in fact a blending of the two, her "structurational model of technology," will provide an accurate and comprehensive research basis: "the reformulation of the technology concept and the structurational model of technology allow a deeper and more dialectical understanding of the interaction between technology and organizations. This understanding provides insight into the limits and opportunities of human choice, technology development and use, and organizational design" (p. 398).

Orlikowski begins her discussion by identifying two "important aspects of the technology concept: ...scope—what is defined as comprising technology, and role—how is the interaction between technology and organizations defined" (p. 398). Although both of these aspects are discussed in the early literature on technology in organizations, they have been viewed as distinct and non-interactive. Orlikowski identifies three "streams of technology research [that] can be distinguished by their definitions of the role played by technology in organizations, reflecting the philosophical opposition between subjective and objective realms that has dominated the social sciences" (p. 399). The early research "assumed technology to be an objective, external force that would have (relatively) deterministic impacts on organizational properties such as structure. In contrast, a later group of researchers focused on the human action aspect of technology, seeing it more as a product of shared interpretations or interventions. The third, and more recent, work on technology has reverted to a 'soft' determinism where technology is posited as an external force having impacts, but where these impacts are moderated by human actors and organizational contexts" (pp. 399-400).

While much of the early and later work focused on how technology and information technology affect organizational structure and performance, it also addressed human aspects such as individual levels of "job satisfaction, task complexity, skill levels, communication effectiveness, and productivity" (p. 400). The most recent perspective, however, is provided by Barley (see Chapter III), "and involves portraying technology as an intervention into the relationship between human agents and organizational structure, which potentially changes it" (p. 402). Working from this base, Orlikowski offers her theory of structuration, which she defines as "a social process that involves the reciprocal interaction of human actors and structural features of organizations...[recognizing] that human actions are enabled and constrained by structures, yet that these structures are the result of previous actions." Orlikowski credits Giddens for establishing the platform for structuration, but points out that Giddens "understood paradigmatically, that is, as a generic concept that is only manifested in the structural properties of social systems (Giddens, 1979, pp. 64-65). Structural properties consist of the rules and resources that human agents use in their everyday interaction. These rules and resources mediate human action, while at the same time they are reaffirmed through being used by human actors" (Orlikowski, 1992, p. 404).

The description of the structurational model of technology is prefaced by a statement that merges the dichotomous views of early research on technology and organizations; Orlikowski understands the recursive relationship that we have with technology and technological development: it is "created and changed by human action, yet it is also used by humans to accomplish some action," and she refers to this relationship as "the duality of technology." Complementing this recursive duality is the notion that "technology is interpretively flexible, hence that the interaction of technology and organizations is a function of the different actors and socio-historical contexts implicated in its development and use" (p. 405).

"The structurational model of technology comprises the following components: (1) human agents—technology designers, users, and decision-makers; (2) technology—material artifacts mediating task execution in the workplace; and (3) institutional properties of organizations, including organizational dimensions such as structural arrangements, business strategies, ideology, culture, control mechanisms, standard operating procedures, division of labor, expertise, communication patterns, as well as environmental pressures such as government regulation, competitive forces, vendor strategies, professional norms, state of knowledge about technology, and socio-economic conditions" (p. 409). These components combine to demonstrate that each of the three stages of research into technology's role in organizations offers "insights into the limitations and contributions of prior conceptualizations of technology," pointing out how each tradition was only "partially correct, but also one-sided." When we view technology from an "imperative" perspective, in which technology is strictly an "objective reality," we are on a deterministic path. However, the "strategic choice school, perceiving technology to be a dynamic, human construction, provides insight into how technology is developed and interpreted, and how through this construction it reflects social interests and motivations. The view of technology as an occasion for structural change provides insight into how the socio-historical context influences the interaction of humans around the use of a technology," and these insights are necessary for the conflation of technology's scope and role, synthesizing the dialectic elements of organizational structure and human action (p. 423).

Fifty Years of Industrial Innovation

Rothwell (1992) provides a comprehensive, chronological review of researchers' findings and assessments of industrial innovation process, which he refers to as "the commercialization of technological change," from the "technology push" and "need pull" of the 1960s to the "systems integration and networking" model of the late 1980s. Referring to this span of years as representing five generations of innovation research, Rothwell quite pointedly makes note of the fact that, "despite more than three decades of empirical research designed to determine 'the characteristics of technically progressive firms', and 'the factors associated with success or failure in innovation', there still exists no precise prescription or recipe for successful innovation" (p. 223).

Rothwell's charting of the generations of industrial innovation provides a lucid foundation for understanding how and why the linear processes of the 1960s gave way to the "coupling model" of the 1970s, only to yield the fourth generation of innovation process, the "'integrated' model of today, [which] marked a shift from perceptions of innovation as a strictly sequential process to innovation perceived as a largely parallel process." The SIN model, or strategic integration and networking model, explains why "innovation is becoming faster; [how] it increasingly involves inter-company networking; and [how] it employs a new electronic toolkit (expert systems and simulation modeling)." During the late 1950s and early 1960s, "technology-push" was the operative model of industrial innovation. Once empirical results of innovations began to be published in the late 1960s, "considerably more emphasis [was placed] on the role of the marketplace in innovation" (p. 221).

Rothwell distills a series of factors common to many of the studies of this early period. These include:

1. Establishing good internal and external communication;
2. Treating innovation as a corporate-wide task: effective functional integration, involving all departments in the project from its earliest stages;
3. Implementing careful planning and project control procedures: committing resources to up-front screening of new projects, regular appraisal of projects;
4. Maintaining efficiency in development work and high-quality production: implementing effective quality control procedures, taking advantage of up-to-date production equipment;

5. Presenting strong market orientation: emphasis on satisfying user-needs, efficient customer linkages, and where possible, involving potential users in the development process;
6. Providing a good technical service to customers, including customer training where appropriate;
7. Ensuring the presence of certain key individuals: effective product champions and technological gatekeepers;
8. Maintaining a high quality of management: dynamic and open-minded managers, ability to attract and retain talented managers and researchers, a commitment to the development of human capital (pp. 223-224).

Rothwell offers another strategic layer to these eight basic and foundational factors of the early period, factors that "outline the essential pre-conditions for sustained corporate innovation to take place." First, management should be overtly open and positive about innovation, supporting an organization's efforts, particularly when it comes to major changes and innovations "to overcoming the barriers and resistance to innovation that often exist in companies." Second, innovation should be a central and strategic factor in an organization's long horizon, "not...an ad hoc process, but one that has direction and purpose." Third, innovation should be tied into long-term planning and "considerations of future market penetration and growth," rather than be one criterion of "short-term return on investment." Fourth, organizations need to be agile, exhibiting "flexibility and responsiveness to change." Fifth, accept risk, as it is inextricably tied to innovation. Knowing that there will be failures puts innovation within realistic boundaries. Sixth, organizations need to be open to innovation and accommodating of entrepreneurship. The notion of an innovation zone within an organization but distinct from its routine operations will promote the "activities of in-house entrepreneurs" (pp. 227-228).

Rothwell goes on to present an opposition of organizational characteristics that belong to the two extreme perspectives at each end of the chronology he covers. The closer one gets to the systems integration and networking model, the more organic, and less rigid, the organization is. Organic organizations maintain characteristics such as a sense of "freedom from rigid rules"; they are more "participative and informal," and the atmosphere is one in which "many views [were] aired and considered" in "face-to-face communication" that is intent on "breaking down departmental barriers." The management emphasis is "on creative interaction and aims," with leadership that is "outward looking" and maintains a "willingness to take on external ideas." Top-down hierarchies are challenged, with information flowing "downwards as well as upwards," and there is a palpable sense of "flexibility with respect to changing needs." This open and positive environment and sensibility is opposed to what Rothwell describes as characteristics of the mechanistic organization. Here, compartmentalization by department and function is the dominant structure. Words that capture the qualities of this type of organization include "hierarchical [and] bureaucratic, [with] many rules and set procedures; formal reporting, long decision chains and slow decision-making; little individual freedom of action; communication via the written word; [and] much information flows upwards; 'directives' flow downwards" (p. 228).

"The fifth generation innovation process," which Rothwell claims was just getting underway in the early 1990s, was fundamentally characterized by "the systems integration and networking model (SIN), [and] represents a somewhat idealized development of the integrated model, but with added features, e.g., much closer strategic integration between collaborating companies." Never at a loss for neologisms, Rothwell offers "electronification" as perhaps "the most significant feature of SIN." By that he means "an increased use of expert systems as a developmental aid, simulation modeling partially replacing physical prototyping, linked supplier/user CAD systems as part of a process of co-development of new products, and closer electronic product design/manufacturing links (integrated CAD/FMS)" (pp. 236-237).

CONCLUSION

Diffusion, as mentioned in the first chapter of this book, is a field that has been considered in thousands of different academic papers in a large number of diverse academic disciplines. As we consider the literature on technology innovation management, there is a clear cluster of articles that relate to diffusion, innovation, and the organization. Most notable concerns regarding direction, culture, and structure are clearly issues of great importance in the innovation management journal literature.

Rothwell (1992) found that the diffusion of industrial innovation can be distilled down into these commonly found factors:

1. Establishment of good internal and external communication,
2. Innovation as a corporate-wide task,
3. Careful planning and project control procedures,
4. Efficiency in development work and high-quality production,
5. Strong market orientation,
6. Good technical support to customers,
7. The presence of certain key individuals, and
8. Overall quality of management.

While Rothwell's list of critical factors is especially helpful to practitioners, several important works have approached this issue from a more theoretic perspective. Orlikowski (1992) advises of the dual nature of technology—as objective reality and socially constructed product. Understanding this dual nature provides insight into the limits and opportunities of technology development and its use and relation to organizational design.

Whether a firm is basically entrepreneurial or conservative in orientation is found to be important (Miller & Freisen, 1982). In the conservative model, product innovation is something that takes place only when absolutely necessary. Conversely, the entrepreneurial model suggests that resources are wasted in the pursuit of excessive novelty, unless executives are alerted to the dangers of too much innovation, and planning processes and structural integration also warn of too much innovation. Another area that has been the focus of much consideration is the identification of systematic factors that influence the type of diversification of an organization (Chaterjee & Wernefelt, 1991). Having considered the issues that relate to organization structure and orientation, the management of technology research that relates more closely to the mainstream diffusion is briefly noted.

It is important to note that the processes associated with the adoption of efficient innovation may also result in the adoption of inefficient innovation (Abrahamson, 1991)—and perhaps the rejection of efficient innovation. Instead of considering diffusion of innovation from the perspective of influence and information flow, Attwell (1992) proposes that know-how and organizational learning are critical for and potential barriers to adoption of innovations. The term 'diffusion' is frequently closely associated to the application of innovation implementation. The consideration of implementation of innovation can be viewed from multiple perspectives. Cooper and Zmud (1990) consider the perspectives of factors research, process research, and political research. Factors research focuses on a variety of individual, organizational, and technological forces that are important to implementation effectiveness. Process research considers social change activities and suggests that implementation success results when there is commitment to change, when the implementation effort exists, when extensive project definition and planning occurs, and when management of the process is guided by organizational change theories. Finally, political research considers how diverse, vested interests of stakeholders affect implementation efforts.

Consideration of the diffusion of patent knowledge through patent citation resulted in interesting insights by Narin and Perry (1987), who found that measuring patents and patent citation data against other performance indicators such as research and development budgets, peer evaluations, the number of research publications produced, and other quantifiable information such as increases in sales and profits is effective for evaluating the overall direction, breadth, and quality of a company's research program. Having considered the diffusion and innovation from an organizational perspective, we now consider the seminal literature that links the management of knowledge and change to the organization.

REFERENCES

Abrahamson, E. (1991). Managerial fads and fashions—the diffusion and rejection of innovations. *Academy of Management Review, 16*(3), 586-612.

Attewell, P. (1992). Technology diffusion and organizational learning—the case of business computing. *Organization Science, 3*(1), 1-19.

Chatterjee, S., & Wernerfelt, B. (1991). The link between resources and type of diversification—theory and evidence. *Strategic Management Journal, 12*(1), 33-48.

Cooper, R.G., & Zmud, R.W. (1990). Information technology implementation research—a technological diffusion approach. *Management Science, 36*(2), 123-139.

Giddens, A. (1979). *Central problems in social theory.* Berkeley, CA: University of California Press.

Miller, D., & Friesen, P.H. (1982). Innovation in conservative and entrepreneurial firms: Two models of strategic momentum. *Strategic Management Journal, 3,* 1-25.

Narin, F.N., & Perry, R. (1987). Patents as indicators of corporate technological strength. *Research Policy, 16*(2-4), 143-155.

Orlikowski, W.J. (1992). The quality of technology—re-thinking the concept of technology in organizations. *Organization Science, 3*(3), 398-427.

Rogers, E. (1983). *Diffusion of innovations.* New York: The Free Press.

Rothwell, R. (1992). Successful industrial innovation—critical factors for the 1990s. *R&D Management, 22*(3), 221-239.

Chapter VI
Knowledge and Change in Organizations

INTRODUCTION

This chapter on the role of knowledge in the operation of organizations consists of two main thrusts: the effects of knowledge (accrual, dissemination, and implementation) on organizational change, and more specifically, the manner and effects of knowledge transfer within and among firms conducting innovative product design and development. We begin with Grant's (1996) view of the importance and processes of knowledge coordination within a firm's administrative hierarchy, and follow with Nonaka and Konno's (1998) concept of *ba*, or a shared space for the creation and emergence of knowledge. Greenwood and Hining's (1996) examination of the role of radical change on their theory of neo-institutionalism focuses on cognition and its relationships to operative procedures and behavior norms, as opposed to the more traditional view of institutionalism, with its fundamental goals of stasis and equilibrium. Leonard-Barton's (1992) article attempts an in-depth view and explanation of how one identifies and exposes organizational capabilities in the face of organizational structures that promote management practices that have the potential to stifle innovation rather than institute and nurture change.

Hagedorn (1993) begins the section on knowledge transfer by investigating the reasons for and processes by which competing firms exchange organizational knowledge, finding a range of distinguishing characteristics between the subject matter and substance of inter-organizational arrangements and the organizational structures and complexities of those firms. Mowery, Oxley, and Silverman (1996) provide a different dimension to this topic through their use of citation patterns of partner firms' patent portfolios as a basis for comparison of allied firms' technological strengths and assets to their output capabilities. Powell, Koput, and Smith-Doerr (1996) take a focused look at the bio-technology industry to support the position that the more complex a field and dispersed its acknowledged experts, the more innovation will be found in the networks created by and for knowledge transfer than in specific individual firms. Olson, Walker, and Ruekert (1995) explain their determination that a firm's coordination structures, regarding the exchange of knowledge, are affected by how innovative a particular product being developed is. Their findings give credence to the positive value that a resource dependency framework for innovation has for understanding and promoting effective interactions among constituencies when developing new products. Sanchez and Mahoney (1996) look at the role of knowledge transfer and the product creation process to explain how concepts of modularity affect both product design and organizational design, as modularity promotes loose couplings among components that are conducive to an increased range of flexibility—obviously an advantage in environments characterized as radically changing.

Leonard-Barton (1988) describes the symbiotic relationship between technological innovation and its adaptation into the organizational environment as one that promotes the view that the initial implementation of a new technology is an outgrowth of the process of invention. Dasgupta and David (1994) employ an economics lens to explain how research is organized in science and technology sectors, pointing out how interrelated and complex their activities are. Both sectors create systems that, on the positive side, can reinforce and help each other, but are unfortunately analogous to fragile machines that have a propensity to become misaligned. We conclude with Moed, Burger, Frankfort, and Van Raan (1985), and their views on knowledge as an instrument of organizational change, and their focus on academic research production rather than the role of knowledge in innovative product and process development or the impacts of knowledge transfer on social and environmental conditions within an organization.

KNOWLEDGE AND ORGANIZATIONAL CHANGE

Knowledge Application vs. Knowledge Creation

Grant (1996) provides a brief literature review of contributions to the research of knowledge-based approaches to understanding "the nature of coordination within the firm...the implications of the knowledge-based view for hierarchy and the location of decision-making authority [and to] determine the boundaries of the firm." Essentially, Grant approaches the firm as "an institution for integrating knowledge," hence the interest in understanding "the characteristics of knowledge and the knowledge requirements of production." Grant's article seeks to explain "the coordination mechanisms through which firms integrate the specialist knowledge of their members" and bases his perspective on premises such as "knowledge is viewed as residing within the individual, and the primary role of the organization is knowledge application rather than knowledge creation." Accepting these premises, knowledge managers will have insight into "the basis of organizational capability, the principles of organization design (in particular, the analysis of hierarchy and the distribution of decision-making authority), and the determinants of the horizontal and vertical boundaries of the firm" (p. 109).

Grant begins his discussion by addressing theories of the firm: "conceptualizations and models of business enterprises, which explain and predict their structure and behaviors." Each theory is "an abstraction of the real-world business enterprise, which is designed to address a particular set of its characteristics and behaviors." Economists apply their understandings of the theory of the firm when "predicting the behavior of firms in external markets." An organizational theorist, alternatively, recognizing the multiple agents and viewpoints contributing to knowledge in an organization, "analyzes the internal structure of the firm and the relationships between its constituent units and departments." The third substratum of Grant's position is the "transaction cost theory" of the firm, "which focused upon the relative efficiency of authority-based organization ('hierarchies') with contract-based organization ('markets')" (p. 109).

Tentative to assert that the "knowledge-based view" has attained the status of a theory, Grant does establish that it "represents a confluence of long-established interests in uncertainty and information with several streams of newer thinking about the firm. To the extent that it focuses upon knowledge as the most strategically important of the firm's resources, it is an outgrowth of the resource-based view [which] recognizes the transferability of a firm's resources and capabilities as a critical determinant of their capacity to confer sustainable competitive advantage." More generally, "knowledge is central to several quite distinct research traditions, notably organizational learning, the management of technology, and managerial cognition." This being the case, the knowledge-based view should be understood as significant to a wider range of concerns than those of "strategic management—strategic choice and competitive advantage...notably the nature of coordination within the firm, organizational structure, the role of management and the allocation of decision-making rights, determinants of firm boundaries, and the theory of innovation" (p. 110).

Grant discusses the issue of transferability, pointing to its importance "not only between firms, but also even more critically, within the firm." Acknowledging some of the standard oppositions such as "distinctions between subjective vs. objective knowledge, implicit or tacit vs. explicit knowledge, personal vs. prepositional knowledge, and procedural vs. declarative knowledge," Grant discusses their importance for organizational theorists in determining how best to maximize both tacit and explicit knowledge. "I identify knowing how with tacit knowledge, and knowing about facts and theories with explicit knowledge. The critical distinction between the two lies in transferability and the mechanisms for transfer across individuals, across space, and across time. Explicit knowledge is revealed by its communication. This ease of communication is its fundamental property. Indeed information has traditionally been viewed by economists as being a public good; once created it can be consumed by additional users at close to zero marginal cost. Tacit knowledge is revealed through its application. If tacit knowledge cannot be codified and can only be observed through its application and acquired through practice, its transfer between people is slow, costly, and uncertain" (p. 111).

Knowledge transfer always involves the willing processes of transmission and receipt of information. Theorists analyze knowledge receipt "in terms of the absorptive capacity of the recipient. At both individual and organizational levels, knowledge absorption depends upon the recipient's ability to add new knowledge to existing knowledge [and the aggregation of knowledge] is greatly enhanced when knowledge can be expressed in terms of common language." Grant defines "the ability to transfer and aggregate knowledge [as] a key determinant of the optimal location of decision-making authority within the firm." Another keyword in the literature of the knowledge-based view is appropriability, which "refers to the ability of the owner of a resource to receive a return equal to the value created by that resource." One important difference between tacit knowledge and explicit knowledge is that tacit knowledge "is not directly appropriable because it cannot be directly transferred: it can be appropriated only through its application to productive activity." As for explicit knowledge, the fact that it is a publicly available resource means that "anyone who acquires it can resell without losing it [and] the mere act of marketing knowledge makes it available to potential buyers" (p. 111).

If one ascribes to the "principle of bounded rationality…[recognizing] that the human brain has limited capacity to acquire, store and process knowledge," one must take into consideration the fact that "efficiency in knowledge production (by which I mean the creation of new knowledge, the acquisition of existing knowledge, and storage of knowledge) requires that individuals specialize in particular areas of knowledge." This is important to "a knowledge-based theory of the firm," given that "the critical input in production and primary source of value is knowledge." In order to benefit economically from knowledge, some theorists posit that "the existence of the firm represents a response to a fundamental asymmetry…knowledge acquisition requires greater specialization than is needed for its utilization. Hence, production requires the coordinated efforts of individual specialists who possess many different types of knowledge." However, some firms fail to coordinate the transfer of knowledge within the firm because "of (a) the immobility of tacit knowledge and (b) the risk of expropriation of explicit knowledge by the potential buyer. Hence, firms exist as institutions for producing goods and services because they can create conditions under which multiple individuals can integrate their specialist knowledge." Therefore, as Spender (1989) has found, "the knowledge-based view of the firm focuses upon the acquisition and creation of organizational knowledge. Thus, Spender defines 'the organization as, in essence, a body of knowledge about the organization's circumstances, resources, causal mechanisms, objectives, attitudes, policies, and so forth'" (Grant, 1996, p. 112).

Grant cautions that "taking the organization as the unit of analysis not only runs the risk of reification, but, by defining rules, procedures, conventions, and norms as knowledge fails to direct attention to the mechanisms through which this 'organizational knowledge' is created through the interactions of individuals, and offers little guidance as to how managers can influence these processes...The knowledge-based view simply focuses upon the costs associated with a specific type of transaction—those involving knowledge." However, "without benefits from specialization there is no need for organizations comprising multiple individuals. Given the efficiency gains of specialization, the fundamental task of organizations is to coordinate the efforts of many specialists. Although widely addressed, organization theory lacks a rigorous integrated, well-developed and widely agreed theory of coordination." Some theorists approach coordination as "the resolution of intra-organizational goal conflict, while the institutional economics literature has been dominated by the problems of the divergence of employee and owner goals causing problems of agency, shirking, and opportunism" (p. 113).

Other theorists approach coordination as being "dependent upon the characteristics of the process technology deployed. Thus, Thompson (1976) identified three types of interdependence, pooled, sequential, and reciprocal, to which Van de Ven, Delbecq, and Koenig (1976) added a fourth, team interdependence." Those promoting a knowledge-based view of the firm suggest that we "perceive interdependence as an element of organizational design and the subject of managerial choice rather than exogenously driven by the prevailing production technology" for we must devise "mechanisms for integrating individuals' specialized knowledge" (Grant, 1996, p. 114).

The knowledge-based view of the firm takes into account "the high costs of consensus decision-making given the difficulties of communicating tacit knowledge. Hence, efficiency in organizations tends to be associated with maximizing the use of rules, routines and other integration mechanisms that economize on communication and knowledge transfer, and reserve problem solving and decision making by teams to unusual, complex, and important tasks" (p. 115). In terms of knowledge shared among team members tasked with decision making, the level of efficient integration is dependent upon "the form of language, shared meaning, or mutual recognition of knowledge domains." Grant indicates that there is a difference between U.S. and Japanese corporations in terms of team-based decision making: "while the hierarchies of Western firms combine the roles of cooperation and coordination, Japanese hierarchies exist primarily to provide the incentive structures to support cooperation, but coordination occurs outside the formal hierarchy" (p. 117).

Keep in mind, however, that when "firms are viewed as institutions for integrating knowledge, a major part of which is tacit and can be exercised only by those who possess it, then hierarchical coordination fails." Moreover, "When managers know only a fraction of what their subordinates know and tacit knowledge cannot be transferred upwards, then coordination by hierarchy is inefficient." Grant reminds us that "bureaucratic systems typically associated with organizational hierarchies rely heavily upon rules and directives, [which are] vehicles for the exercise of authority...and [are applied] top down. In the knowledge-based firm, rules and directives exist to facilitate knowledge integration; their source is specialist expertise, which is distributed throughout the organization." Ultimately, to the extent that 'higher-level decisions' are dependent upon immobile 'lower-level' knowledge, hierarchy impoverishes the quality of higher-level decisions" (p. 118). In the end, "the primary role of the firm [is to integrate] the specialist knowledge resident in individuals into goods and services. The primary task of management is establishing the coordination necessary for this knowledge integration" (p. 120).

The Concept of *Ba*

Nonaka and Konno (1998) are concerned with explaining how the Japanese philosophical concept of *ba* is applicable to knowledge creation in innovative environments. *Ba* "can be thought of as a shared space for emerging relationships," with the space being physical, virtual, mental, "or any combination of them." Human interaction takes place in many ways and in each of these environments, but the difference between what we consider to be normal, everyday interaction and *ba* is the outcome of the interaction being knowledge creation in the latter. Not only is *ba* "a platform for advancing individual and/or collective knowledge," it is the space where "a transcendental perspective integrates all information needed...a context which harbors meaning...a shared space that serves as a foundation for knowledge creation" (p. 40).

By combining our experience with the knowledge created in this space and reflecting on it, knowledge becomes more than information. In fact, "information resides in media and networks. It is tangible. In contrast, knowledge resides in *ba*. It is intangible." Nonaka and Konno describe four types of *ba*, all of which to different degrees allow "the individual [to realize] himself as part of the environment on which his life depends." By participating in *ba*, one can "transcend one's own limited perspective of boundary. This exploration is necessary in order to profit from the 'magic synthesis' of rationality and intuition that produces creativity" (p. 41).

The authors differentiate between explicit and tacit knowledge, with the former being that which "can be expressed in words and numbers and shared in the form of data, scientific formulae, specifications, manuals, and the like," all promoting information being "transmitted between individuals formally and systematically." As the Japanese understand knowledge to be primarily tacit, to understand Nonaka and Konno's model of knowledge creation, one must accept their view that tacit knowledge, "something not easily visible and expressible," is a necessary condition for creative thought. "Tacit knowledge is deeply rooted in an individuals' actions and experience as well as in the ideals, values, or emotions he or she embraces." Tacit knowledge can be thought of as having two dimensions: a technical dimension "which encompasses the kind of informal personal skills or crafts often referred to as 'know-how,' [and] the cognitive dimension, [consisting] of beliefs, ideals, values, schemata, and mental models which are deeply ingrained in us and which we often take for granted" (p. 42).

Nonaka and Konno present their SECI model of knowledge creation, with the acronym representing socialization, externalization, combination, and internalization. Finding that "knowledge creation is a spiraling process of interactions between explicit and tacit knowledge," the authors suggest that the "combination of the two categories makes it possible to conceptualize [the] four conversion patterns" of the SECI model. "Socialization involves the sharing of tacit knowledge between individuals. [It] is exchanged through joint activities...rather than through written or verbal instructions" (p. 42). Socialization "involves capturing knowledge through physical proximity [and] disseminating tacit knowledge...The process of transferring one's ideas or images directly to colleagues or subordinates means to share personal knowledge and create a common space—or *ba*" (pp. 42-43).

"Externalization requires the expression of tacit knowledge and its translation into comprehensible forms that can be understood by others...During the externalization stage of the knowledge-creation process, an individual commits to the group, and thus becomes one with the group." There are two factors that support externalization. First "the conversion of tacit into explicit knowledge [which] involves techniques that help to express one's ideas or images as words, concepts, figurative language...and visuals." Second, "translating the tacit knowledge of customers or experts into readily understandable forms" (p. 43). "Combination involves the conversion of explicit knowledge into more complex sets of explicit knowledge, [with] key issues [being] communication and diffusion processes and the systemization of knowledge" (p. 44). There are three phases to combination, the first being "capturing and integrating new explicit knowledge." Second is "the dissemination of explicit knowledge...based on the process of transferring this form of knowledge directly [via] presentations and meetings...Third, the editing or processing of explicit knowledge makes it more usable." Internalization refers to "the conversion of explicit knowledge into the organization's tacit knowledge. This requires the individual to identify the knowledge relevant for one's self within the organizational knowledge." Internalization can be thought of as relying on two dimensions. "First, explicit knowledge has to be embodied in action and practice...Second, there is a process of embodying the explicit knowledge by using simulations or experiments to trigger learning by doing processes" (p. 45).

Corresponding to each stage of the SECI model are four types of *ba*. The first, originating *ba*, is related to socialization; it "is the world where individuals share feelings, emotions, experiences, and mental models...Originating *ba* is the primary *ba* from which the knowledge-creation process begins...Physical, face-to-face experiences are the key to conversion and transfer of tacit knowledge" (p. 46). Interacting *ba* refers to "the place where tacit knowledge is made explicit, thus it represents the externalization process. Dialogue is key for such conversions; and the extensive use of metaphors is one of the conversion skills required" (p. 47). Cyber *ba*, "a place of interaction in a virtual world instead of real space and time...represents the combination phase." By combining "new explicit knowledge with existing information...knowledge [can be generated and systematized] throughout the organization." Finally, exercising *ba* "supports the internalization phase. Exercising *ba* facilitates the conversion of explicit to tacit knowledge...Exercising *ba* synthesizes Nishida's world [the originating Japanese philosophy] and the Cartesian world through action, while interacting *ba* achieves this through thought" (p. 47).

Nonaka and Konno give three examples of companies that have created and employed *ba* as an organizational strategy. "Sharp created *ba* for organic concentration outside of the existing business organization...Toshiba established an internal agent with the function of achieving organic concentration within the existing organization...Maekawa Seisakusho has been built with organic concentration of resources geared to market niches" (p. 48). For each, the "organic concentration of knowledge assets in *ba* involves not a consumption process of resources, but an ecological process with a cyclical cultivation of resources." The authors find that "knowledge is manageable only insofar as leaders embrace and foster the dynamism of knowledge creation...Their task is to manage for knowledge *emergence*...The management of knowledge as a static stock disregards the essential dynamism of knowledge creation...knowledge needs to be nurtured, supported, enhanced, and cared for. Thinking in terms of systems and ecologies can help provide for the creation of platforms and cultures where knowledge can freely emerge" (pp. 53-54).

Traditional and Neo-Institutional Theories of Organizational Change

Greenwood and Hining (1996) establish "a framework for understanding organizational change from the perspective of neo-institutional theory" (p. 1022). The benefit of such a perspective is that it offers an "account of change, first, by providing a convincing definition of radical (as opposed to convergent) change, and, second, by signaling the contextual dynamics that precipitate the need for organizational adaptation" (p. 1023). Greenwood and Hining also "provide an explanation of both the incidence of radical change and of the extent to which such change is achieved through evolutionary or revolutionary pacing" by offering "a more complete account for understanding organizational interpretations of, and responses to, contextual pressures, by stressing the political dynamics of intra-organizational behavior and the normative embeddedness of organizations within their contexts" (pp. 1023-1024). The authors seek to uncover "the processes by which individual organizations retain, adopt, and discard templates for organizing, given the institutionalized nature of organizational fields" (p. 1022).

Quoting Scott (1994), who sees "convergent developments among the approaches of many analysts as they recognize the importance of meaning systems, symbolic elements, regulatory processes, and governance systems" (p. 78), Greenwood and Hining (1996) identify "this convergence around multiple themes, the coming together of the old and the new institutionalism [as] neo-institutionalism." One shortcoming of institutional theory is that it "is not usually regarded as a theory of organizational change, but usually as an explanation of the similarity ('isomorphism') and stability of organizational arrangements in a given population or field of organizations" (p. 1023).

Greenwood and Hining identify three themes to establish their explanation of institutional theory: First is the idea that "a major source of organizational resistance to change derives from the normative embeddedness of an organization within its institutional context." Second is the suggestion that "the incidence of radical change, and the pace by which such change occurs, will vary across institutional sectors because of differences in the structures of institutional sectors, in particular in the extents to which sectors are tightly coupled and insulated from ideas practiced in other sectors." Third is the proposition that "both the incidence of radical change and the pace by which such change occurs will vary within sectors because organizations vary in their internal organizational dynamics" (p. 1023).

The authors define radical organizational change as a breaking away from an existing organizational orientation that is transformative; convergent change is the process of "fine-tuning the existing orientation...Revolutionary and evolutionary changes are defined by the scale and pace of upheaval and adjustment. Whereas evolutionary change occurs slowly and gradually, revolutionary change happens swiftly and affects virtually all parts of the organization simultaneously" (p. 1024). If behaviors derive from ideas, values, and beliefs that originate in an organizational context, "organizations must accommodate institutional expectations, even though these expectations may have little to do with technical...performance accomplishment." We look to institutional theory to show us "how organizational behaviors are responses not solely to market pressures, but also to institutional pressures (e.g., pressures from regulatory agencies, such as the state and the professions, and pressures from general social expectations and the actions of leading organizations)." When we place emphasis on values that are "suggested [by] the configuration or pattern of an organization's structures and systems," we are providing institutional theory with "an interpretive scheme" (p. 1025). Greenwood and Hining also offer a second perspective on institutional theory by considering "the structure of the institutional context (i.e., the extent of tight coupling and the extent of sectoral permeability)" in which "sectors usually have been perceived as having clearly legitimated organizational templates and highly articulated mechanisms (the state, professional associations, regulatory agencies, and leading organizations) for transmitting those templates to organizations within the sector" (p. 1029).

There is a difference between "old institutionalism," which "emphasizes the details of an organization's interactions with its environment over time and pays attention to the beliefs and actions of those who have the power to define directions and interests," and new institutionalism, which "emphasizes the regulative, the normative, and the cognitive. In this case, rather than values and moral frames, it is cognition that is important. As Meyer and Rowan (1977, p. 341) put it, 'normative obligations enter into social life primarily as facts.' The key units of analysis are organizations-in-sectors and their relation to societal institutions" (Greenwood & Hining, 1996, pp. 1031-1032). It is the shift from a values-based perspective to a cognitive-based platform for change, one that pays special attention to how "complex organizations handle growth and/or contextual complexity by differentiation into groups, each of which is focused on specialized tasks" and the "process of specialization [that] leads to significant differences between groups in terms of structural arrangements and orientation" that demarcates the shift from "old-" to "neo-institutionalism" (p. 1033).

Greenwood and Hining recognize that "organizations are arenas in which coalitions with different interests and capacities for influence vie for dominance" (p. 1035), and this undergirds their effort "to show how the external processes of deinstitutionalization have to be understood (organizations-in-sectors) together with the internal dynamics of interpretation, adoption, and rejection by the individual organization" (p. 1041). Combined with the suggestion that "the understanding of radical change requires more than an analysis of the institutional arena or sector"

(p. 1042), we can begin to see that "change is about understanding variations in response to the same pressures, which can only be done by analyzing the features of organizations that produce adoption and diffusion rather than resistance and inertia" (p. 1041).

Knowledge and Organizational Core Capabilities

Leonard-Barton (1992) seeks to provide an overview of the literature of the nature of a firm's core capabilities in order to put the concept into opposition with another critically strategic concept—core rigidities. From Leonard-Barton's "knowledge-based" perspective, understanding a firm's core capabilities is important for the "development of new products and processes." Her primary question is: how can a deeper understanding of "core capabilities and detailed evidence about their symbiotic relationship with development projects" help management strategists tackle the capability/rigidity paradox: how can new product development occurring within organizational structures in which "observed management tactics" that can be considered hindrances to "potential...product/process development projects" be turned around "to stimulate change?" (p. 111).

Core capabilities, defined as "the knowledge set that distinguishes and provides a competitive advantage" (p. 112), are "traditionally treated as clusters of distinct technical systems, skills, and managerial systems, [and] these dimensions of capabilities are deeply rooted in values, which constitute an often overlooked but critical 4th dimension." Embedded in these core capabilities are core rigidities, management strategies, policies, and technological systems "that inhibit innovation...Managers of new product and process development projects thus face a paradox: how to take advantage of core capabilities without being hampered by their dysfunctional flip side" (p. 111).

Whether one refers to core capabilities that "differentiate a company strategically" as "distinctive competences," "core organizational competencies," "firm-specific competence," "resource deployments," or "an invisible asset," as they have been called in the literature since the late 1970s, one area of contention among researchers has been whether core capabilities should be considered strategic or tactical assets. Some find that "industry-specific capabilities increased the likelihood a firm could exploit a new technology within that industry," while others contend that "effective competition is based less on strategic leaps than on incremental innovation that exploits carefully developed capabilities." What a firm needs to keep in mind is how core capabilities "may lead to 'incumbent inertia' in the face of environmental changes." Leonard-Barton's warning: "At any given point in a corporation's history, core capabilities are evolving, and corporate survival depends upon successfully managing that evolution" (p. 112). This exposes the tension between institutionalization of processes and policies and the ability to adapt to change, both internal and external.

Leonard-Barton offers four dimensions to the content of a knowledge set that "distinguishes and provides a competitive advantage: (1) employee knowledge and skills are embedded in (2) technical systems. The processes of knowledge creation and control are guided by (3) managerial systems. The fourth dimension is (4) the values and norms associated with the various types of embodied and embedded knowledge and with the processes of knowledge creation and control" (p. 112).

This content accrues through organizational capabilities. For example, "knowledge and skills embodied in people" are tied closely to "new product development. This knowledge/skills dimension encompasses both firm-specific techniques and scientific understanding...knowledge embedded in technical systems...results from years of accumulating, codifying and structuring the tacit knowledge in peoples' heads...managerial systems...represent formal and informal ways of creating knowledge...and of controlling knowledge. [These] dimensions [precipitate] the value assigned within the company to the content and structure of knowledge [and the] means of collecting knowledge...and controlling knowledge" (p. 113).

Two dimensions of the value ascribed to "knowledge creation and content" are "the degree to which project members are empowered and the status assigned various disciplines on the project team." Leonard-Barton defines empowerment as "the belief in the potential of every individual to contribute meaningfully to the task at hand and the relinquishment by organizational authority figures to that individual of responsibility for that contribution." Actualizing this belief occurs when an organization, "generally recognized for certain core capabilities attracts, holds, and motivates talented people who value the knowledge base underlying that capability and join up for the challenges, the camaraderie with competent peers, [and] the status associated with the skills of the dominant discipline or function" (p. 115). There is a downside to the instantiation of these values. Even though "projects derive enormous support from core capabilities [and] such capabilities continually spawn new products and processes because so much creative power is focused on identifying new opportunities to apply the accumulated knowledge base...these same capabilities can also prove dysfunctional for product and process development." When this happens—when "values, skills, managerial systems, a technical systems that served the company well in the past and may still be wholly appropriate for some projects or parts of projects, are experienced by others as core rigidities—inappropriate sets of knowledge," it has a destabilizing effect on new product development, specifically, and the organization, generally: "These deeply embedded knowledge sets actively create problems" (p. 116). After all, to separate the personal investment of employee knowledge, good will, and effort—what Leonard-Barton considers "a psychological contract with the corporation"—into new product development as a result of the "intractable" nature of some management systems (core rigidities), is to hinder or squelch the innovation completely. "Highly skilled people are understandably reluctant to apply their abilities to project tasks that are undervalued, lest that negative assessment of the importance of the task contaminate perceptions of their personal abilities...the very same values, norms and attitudes that support a core capability and thus enable development can also constrain it" (p. 117).

Leonard-Barton identifies several human manifestations of a company's bequeathing of high status on innovative projects. These include "who travels to whom," "self-fulfilling expectations," and "unequal credibility and wrong language." Unfortunately, as "dozens of controlled experiments manipulating unconscious interpersonal expectations have demonstrated, biases can have a 'Pygmalion effect': person A's expectations about the behavior of person B affect B's actual performance [that] can be dangerously self-fulfilling" (p. 118). This is why it is critical for project managers to understand the nature of the core capabilities/core rigidities paradox: how many different forms does each take, and what is the nature of their differences? "The more dimensions represented, the greater the misalignment potentially experienced between project and capability" (p. 118).

The dimensions of project value vary, but can be understood as connected in terms of how easily technical systems, managerial systems, individual skill sets and an individual's values change; in other words, how much resistance core rigidities exert when the process of innovative product development is in play. The "dimensions are increasingly less tangible, less visible and less explicitly codified." The easiest to alter is the "technical systems dimension" because "such systems are local to particular departments." Managerial systems are more complex in scope, as they "reach across more subunits than technical systems, requiring acceptance by more people." When it comes to the "skills and knowledge content dimension," change is even more difficult "because skills are built over time and many remain tacit, i.e. un-codified and in employees' heads" (p. 119). When we consider that "the value embodied in a core capability is the dimension least susceptible to change," however, we begin to realize the importance of project managers addressing the capabilities/rigidities paradox effectively. In her study, Leonard-Barton found that "managers handled the paradox in one of four ways: (1) abandonment; (2) recidivism, i.e., return to core capabilities; (3) reorientation; and (4) isolation." These management approaches can "pave the way for organizational change by highlighting core rigidities and introducing new capabilities" (p. 119).

If one considers core rigidities from an evolutionary perspective, one can see that "small departures from tradition in organizations [can provide] a foundation in experience to inspire eventual large changes." One way of instigating change is to introduce "new capabilities along any of the four dimensions. However, for a capability to become core, all four dimensions must be addressed. A core capability is an interconnected set of knowledge collections—a tightly coupled system" (p. 119). In order for new technical systems to provide advantage, new skills must be developed and deployed. As in any organic system, "New skills atrophy or flee the corporation if the technical systems are inadequate, and/or if the managerial systems such as training are incompatible. New values will not take root if associated behaviors are not rewarded" (p. 120).

KNOWLEDGE TRANSFER

Inter-Firm Knowledge Exchange

Increased competition for market share of so many products and services worldwide has led some researchers, such as Hagedorn (1993), to seek answers to questions such as: "Why [do] companies cooperate in their efforts to innovate?" How do we define "the domain of both vertical and horizontal inter-firm relationships as well as in short-term and in long-term perspective of cooperation?" What are the "motives that play a role in inter-firm strategic technology partnering, given certain sectoral differences that obviously could influence the motivation of firms to collaborate?" (pp. 371, 374). What distinguishes Hagedorn's article from much of the previous research in organizational knowledge is the "attention...paid to both sectoral differences in the motivation for partnerships as well as to contrasts in interorganizational features of technology cooperation." His "analysis reveals some major differences regarding the research orientation of contractual arrangements and organizationally complex alliances" (p. 371).

In the normal course of events, the boundaries of any particular firm are "defined in terms of vertical relationships of economic exchange from one company to another. Technology cooperation frequently surpasses this particular arena of economic exchange and enters into a field of relatively long-term strategic considerations regarding lateral relationships between companies." Researchers viewing organizational boundaries from the perspective of transaction cost economics suggest that "inter-firm partnering as an economic phenomenon in between market transactions and hierarchies" is quite different from traditional vertical relationships (p. 371).

Inter-organizational cooperation is often spurred by "the reduction, minimizing and sharing of...uncertainty, which is inherent to performing R&D. Many studies refer to the reduction of risk in R&D as a major motive for shared activities; we, however, suggest it is more appropriate to think of this sharing of R&D in terms of reduction of uncertainty." While risk refers to "the probability of occurrence of an event with a given probability distribution of the size of the event," uncertainty is "associated with the unknown likelihood of an event when there is no probability distribution" (p. 372). Hagedorn cites the strength of the unknown to motivate firms "to combine their efforts in order to create economies of scale and/or scope that will facilitate their search processes to expand to a wider field of research activities or expand their competence" (pp. 372-373). It is the "complexity and interrelatedness of different fields of technology and their efforts to gain time and reduce uncertainty in joint undertakings during a period of growing technological intricacy" that compels competing firms to transcend vertical boundaries and form strategic R&D alliances (p. 378).

Hagedorn identifies two basic categories of ways that "inter-firm cooperative agreements and motives across a broad spectrum of alternative [create] incentives to collaborate: market and technology-related motives." Because technology-related motives are so prevalent in high-tech sectors, attention to sectoral differences must be understood. As strategies for lateral relationships between companies become more complex, driven in large part by market competitiveness, "complex agreements with equity investments and shareholder control partners increase their governance over their strategic alliances." An alternative strategy is to engage in "contractual arrangements [that] demand less control through administration and supervision" than inter-firm agreements, as they are "less complex regarding their span of objectives." Hagedorn (1993) suggests that "companies appear to prefer this mode of strategic technology partnering for agreements with a one-dimensional goal, strongly biased in favor of applied research cooperation" (pp. 381-382).

Measuring the Value of Knowledge Partnerships

Like Hagedorn (1993), inter-firm knowledge transfer within strategic alliances is the subject of Mowery et al. (1996). The authors use citation patterns of partner firms' patent portfolios to "measure changes in the extent to which their technological resources 'overlap' with their partners' technological portfolios as a result of participation in an alliance" (p. 78). While their "empirical investigation focuses on transfer of technological capabilities among alliance partners," they are particularly interested "in how collaboration changes the relationship between a firm's technological portfolio and those of its alliance partner(s)" (p. 79). Their research questions concern whether "equity arrangements promote greater knowledge transfer...'Absorptive capacity' helps explain the extent of technological capability transfer [and whether there are discernable limits] to the 'capabilities acquisition' view of strategic alliances." Their findings include "that alliance activity can promote increased specialization [and] that the capabilities of partner firms become more divergent in a substantial subset of alliances" (p. 77).

Mowery et al.'s interest in strategic alliances as an element in a firm's competitive strategies derives from two concepts prevalent in the literature: the "resource-based view of the firm, [which] describes the business enterprise as a collection of sticky and difficult-to-imitate resources," and the notion of "dynamic capabilities, [which] emphasizes the importance of change in the capabilities underpinning these resources…focusing in particular on the development, more than the exploitation, of firm-specific resources." Organizational learning is one "dynamic capability" that can serve as a firm's strategic pathway to new acquisitions. Concomitantly but from another perspective, knowledge-based views of the firm "focus on knowledge as a key competitive asset, and emphasize the capacity of the firm to integrate tacit knowledge" (p. 77). Strategic alliances "have advantages over conventional contracts or markets…because firm-specific technological capabilities frequently are based on tacit knowledge and are subject to considerable uncertainty concerning their characteristics and performance" (p. 79).

Mowery et al. find that "there are limits to the 'capabilities acquisition' view of alliances." Not only can "alliance activity…lead to increased specialization, as firms access others' capabilities…the capabilities of partner firms become more divergent in a substantial subset of alliances" (p. 78). Their perception of the established literature on inter-firm knowledge transfers in alliances supports the opinion that "equity joint ventures appear to be more effective conduits for the transfer of complex capabilities than are contract-based alliances such as licensing agreements." Moreover, "lower levels of transfer occur [more] in unilateral contracts than in bilateral non-equity arrangements, [suggesting] that the structure and content of alliances are jointly determined, and that alliances nearer the 'hierarchy' end of the 'market- hierarchy' continuum outperform alternatives in supporting inter-firm learning." The authors' data supports the Cohen and Levinthal (1990) position on "the importance of 'absorptive capacity' in the acquisition of capabilities through alliances and bolsters the argument that experience in related technological areas is an important determinant of absorptive capacity." Ultimately, "a firm's ability to absorb capabilities from its alliance partner depends on the pre-alliance relationship between the two firms' patent portfolios, [which is] consistent with Cohen and Levinthal's characterization of absorptive capacity as a quality that is both firm-specific and path-dependent" (Mowery et al., 1996, p. 89).

Sector-Specific Advantages to Knowledge Transfer

Powell et al. (1996) approach the subject of organizational knowledge and inter-firm alliances by viewing "forms of collaboration undertaken by dedicated biotechnology firms and [assessing] the contribution of cooperative ventures to organizational learning." Their efforts are focused on identifying the network structure of what was, in 1996, an emerging field, biotechnology, and "[explaining] the purposes served by the extensive connections that typify the field" (p. 117).

Powell et al. expand on other researchers whose lens is knowledge-based to "argue...that when the knowledge base of an industry is both complex and expanding and the sources of expertise are widely dispersed, the locus of innovation will be found in networks of learning, rather than in individual firms." Their contribution is the articulation of "a network approach to organizational learning and...firm-level, longitudinal hypotheses that link research and development alliances, experience with managing inter-firm relationships, network position, rates of growth, and portfolios of collaborative activities." With data drawn from "a sample of dedicated biotechnology firms in the years 1990-1994," the authors support their hypothesis that "pooled, within-firm, time series analyses support a learning view and have broad implications for future theoretical and empirical research on organizational networks and strategic alliances" (p. 116).

Why do research firms collaborate? Powell et al. posit that it is a response to "some combination of risk sharing, obtaining access to new markets and technologies, speeding products to market, and pooling complementary skills" (p. 116). Considering the rapid rate of technological development in areas such as biotechnology, "research breakthroughs are so broadly distributed that no single firm has all the internal capabilities necessary for success [making] new technologies...both a stimulus to and the focus of a variety of cooperative efforts that seek to reduce the inherent uncertainties associated with novel products or markets" (p. 117). Moreover, if one takes a social constructionist position, "what is learned is profoundly linked to the conditions under which it is learned. Knowledge creation occurs in the context of a community, one that is fluid and evolving rather than tightly bound or static." This position minimizes the positive effects that a "canonical formal organization, with its bureaucratic rigidities," can have in terms of learning, knowledge transfer, and the innovations deriving from them. "Sources of innovation do not reside exclusively inside firms; instead, they are commonly found in the interstices between firms, universities, research laboratories, suppliers, and customers" (p. 118).

Another perspective to adopt is one that considers "the differences between exploration and exploitation in organizational learning." Powell et al. cite March's argument "that the 'essence of exploitation is the refinement and extension of existing competencies, technologies and paradigms. The essence of exploration is experimentation with new alternatives'" (March, 1991, p. 85). Exploitation generates predictable returns, while the returns from exploration are much more uncertain. "Exploration is costly, often unfruitful," but a winning strategy nonetheless (Powell et al., 1996, p. 118).

Powell et al. present a computing network analogy to describe the circumstance "that when knowledge is broadly distributed and brings a competitive advantage, the locus of innovation is found in a network of inter-organizational relationships." Currency depends to a great extent on existing as an active player within that network, as passive "recipients of new knowledge are less likely to appreciate its value or to be able to respond rapidly." They recommend that firms in "industries in which know-how is critical...must be expert at both in-house research and cooperative research with such external partners as university scientists, research hospitals, and skilled competitors;" the authors augment their analogy by comparing the "network analog to Cohen and Levinthal's (1989, 1990) concept of 'absorptive capacity.' A firm with a greater capacity to learn is adept at both internal and external R&D, thus enabling it to contribute more to collaboration as well as learn more extensively from such participation." Most importantly, "a network serves as a locus of innovation because it provides timely access to knowledge and resources that are otherwise unavailable, while also testing internal expertise and learning capabilities" (Powell et al., 1996, p. 119).

Powell et al. conclude by reminding readers that "standard organizational characteristics, such as age and size, appear to be ancillary in accounting for patterns of collaboration. Neither growth nor age reduced the propensity to engage in external relationships. Instead, age, per se, proved unimportant in the context of network experience, and size was an outcome rather than a determinant of partnerships." Indeed, "firms without ties are becoming increasingly rare; the modal firm has multiple partnerships. Perhaps our most interesting descriptive result is that the field is becoming more tightly connected not in spite of, but because of a marked increase in the number of partners involved in alliances" (p. 143).

Knowledge Transfer and Organizational Structure

Olson et al. (1995) discuss the role of participatory teams in successful new product development as one that is beneficial in terms of the sharing of expertise and the dismantling of bureaucratic organizational hierarchies. Their study of 45 products from a diverse field of manufacturers does not suggest that traditional stove-piping of functions and divisions is necessarily inefficient or even passé, as there is no "one type of coordinating structure [that] is likely to be uniformly successful in delivering more creative new products, cutting development time, and improving new product success in the marketplace across all kinds of development projects." In fact, "the degree of innovativeness or newness of the product being developed is an important moderator of the impact of different coordination structures on the development process and its outcomes" (p. 48). The authors parse the different moderating elements in terms of their potential contributions to innovative product development.

Given the increasing heterogeneity of design and development groups, some involving the expertise of marketers and sales personnel, Olson et al. offer insight into how such "cross-functional interactions can be structured and coordinated in a variety of ways, from bureaucratic approaches to more decentralized participatory mechanisms." Relying on "resource dependency theory, which suggests that more participative structures are likely to improve the effectiveness and timeliness of the development process when the product developed is truly new and innovative," the authors argue that although "more bureaucratic structures may produce better outcomes on less innovative projects, such as those involving line extensions or product improvements," their findings point to the importance of understanding and exploiting "the fit between the newness of the product concept and the participativeness of the coordination mechanism used." By doing so, the development process stands to yield better outcomes "in terms of (1) objective measures of product and team performance, (2) the attitudes of team members toward the process, and (3) the efficiency and timeliness of the new product development process" (p. 48).

Typically, organizational structure can be characterized as reflective of "a division of work according to functional specialties such as marketing, finance, production, and research and development (R&D)." Such a structure is beneficial when one seeks "efficiencies within each specialty" of a particular firm, but "they also give rise to a need for cross-functional interaction and coordination" that relies on "lateral linkage devices or structural coordination mechanisms to connect relatively autonomous functional units" to support coordination (p. 49). Previous literature suggests that there are "similar sets of lateral linkage mechanisms that organizations use to coordinate inter-functional interactions across the full spectrum of organizational activities, including the new product development process," and Olson et al. point to the following as prominent moderators:

- **Bureaucratic control/hierarchical directives:** These are the "most formalized and centralized—and the least participative—mechanism [which] relies on standard operating procedures and the oversight of a high-level general manager to coordinate activities across functions."
- **Individual liaisons:** This is the role of "individuals within one or more functional departments [who] are assigned to communicate directly with their counterparts in other departments, thus supplementing some of the vertical communication flow found in bureaucracies." Even though they may "carry no formal authority to make decisions aimed at resolving inter-functional conflicts...they can often wield informal influence by virtue of their centrality within communication networks that cross functional boundaries."
- **The temporary task force:** This "represents an institutionalization of the repetitive interaction among liaison individuals within the context of a specific

project. Because task force members represent various functions and interact directly, this is a more participative and less formalized mechanism than those above."

- **An integrated manager:** This represents "an additional management position [that] is superimposed on the functional structure."
- **Matrix structures:** These are structures, in which activities are structured "according to product or market focus as well as by function" (p. 49).

More recently, design teams and design centers have "gained popularity as organizations have searched for ways to improve the timeliness and effectiveness of their product development efforts within ever more rapidly changing environments. A design team is similar to a temporary task force and a matrix structure in that it [brings] together a set of functional specialists to work on a specific new product development project. Unlike [bureaucratic controls and individual liaisons], however, such teams tend to be more self-governing and have greater authority to choose their own internal leader(s), establish their own operating procedures, and resolve conflicts through consensual group processes" (pp. 49-50). Olson et al.'s results "confirm the usefulness of the resource dependency framework in understanding functional interactions within the context of new product development. They indicate that when a firm and its potential customers are relatively unfamiliar and have little previous experience with a new product concept, the functional tasks involved in developing the concept and bringing it to market are more difficult and challenging than when the project involves a more straightforward modification or extension of an existing line" (p. 59).

Modularity's Role in Knowledge Transfer

Sanchez and Mahoney (1996), like Olson et al., turn their attention to the role of knowledge transfer and the product creation process to explain how concepts of modularity affect both product design and organizational design. An understanding of how "advanced technological knowledge about component interactions can be used to fully specify and standardize the component interfaces that make up a modular product architecture, creating a nearly independent system of 'loosely coupled' components," is key to "improving a firm's strategic flexibility to respond advantageously to a changing environment." Sanchez and Mahoney describe modular organizational structure as effectively facilitating "specific forms of 'coordinated self-organizing processes.'" What each component of an organization's hierarchy does in the product design process is equaled in importance by an understanding of the "interrelationships of product design, organization design, processes for learning and managing knowledge, and competitive strategy." Sanchez and Mahoney consider these components to be "nearly decomposable systems" in themselves. If an organization has "the ability of standardized interfaces between components in a product design to embed coordination of product development processes," it will create "hierarchical coordination" that does not contain "the need to continually exercise authority—enabling effective coordination of processes without the tight coupling of organizational structures" (p. 63).

Previous research, such as that of Daft and Lewin (1993), establishes that "modular organization" is derived from "the need for flexible, learning organizations that continuously change and solve problems through interconnected coordinated self-organizing processes" (p. i). Without the need for "overt exercise of managerial authority to achieve coordination of development processes," a need obviated by "standardized component interfaces in a modular product architecture [which] provide a form of embedded coordination...concurrent and autonomous development of components by loosely coupled organization structures" is possible (p. 64).

For Sanchez and Mahoney, a complex system is one that "consists of parts that interact and are interdependent to some degree." The system's complexity is an outcome of its hierarchy, in this case, one in which the whole is "essentially composed of interrelated subsystems that in turn have their own subsystems." Such a view of organizational hierarchy demands a "more general conception of 'hierarchy' than that usually invoked in organizational economics and strategic management...where hierarchy typically denotes subordination to an authority relationship." Sanchez and Mahoney use the term as referring to "a decomposition of a complex system into a structured ordering of successive sets of subsystems...a partitioning into relationships that collectively define the parts of any whole" (p. 64).

When the end products of a component development process are "partitioned into tasks that can be performed autonomously and concurrently by a loosely coupled structure of development organizations," the organization can be more flexible, as "the information structure provided by the standardized component interface specifications of modular product architecture provides a means to embed coordination of loosely coupled component development processes" (p. 66). Sanchez and Mahoney use a computer programming analogy to describe product development projects occurring in loosely coupled organizational structures: such projects "can be thought of as 'programmed' innovation in which firms create new products by applying existing knowledge and creating new knowledge about components and their interactions" (p. 68). The object-oriented programming concept of "classes" is particularly apt in this circumstance. Instead of following the traditional model in which there is a "sequential staging of design and development tasks...After defining the product concept, design and development tasks are sequenced so that technology and component development tasks with the greatest need for new knowledge and with the greatest impact on other component design and development tasks are undertaken first" (p. 68).

In a product development environment comprising "overlapping development stages," there is "greater sharing of current information through processes of overlapping problem solving that link closely interrelated component design and development tasks." Such environments facilitate learning "through the development of individual components," as the "stable information structure of fully specified product architecture" results in an avoidance of "learning inefficiencies due to breakdowns, losses, and delays in information flows between component developments activities" (p. 70). Sanchez and Mahoney provide insight to "the potential for intentionally decomposing complex products and organizational phenomena into loosely coupled subsystems," a fundamental step toward understanding "the structure and dynamics of changing product markets and evolving organizational forms" (p. 73).

Knowledge Transfer, Innovation, and Organizational Change

Leonard-Barton (1988) offers readers insights into "processes of initial technology implementation" by taking "a deliberately cross-disciplinary stance in suggesting that initial implementation of technical innovations is best viewed as a process of mutual adaptation, i.e., the re-invention of the technology and the simultaneous adaptation of the organization" (p. 253). In other words, there is a symbiotic relationship between technological innovation and its adaptation into the organizational environment, which causes students of technology transfer to question the division believed by many to exist between innovation and implementation. Leonard-Barton considers the initial implementation of a new technology to be "an extension of the invention process." While it would be convenient in many ways for "the predictable realization of a preprogrammed plan [to take form in the] implementation [as] a dynamic process of mutual adaptation between the technology and its environment," in effect obviating all unintended consequences, Leonard-Barton suggests that Van de Ven (1986) should be heeded: "Innovations not only adapt to existing organizational and industrial arrangements, but they also transform the structure and practice of these environments" (p. 591). Leonard-Barton views such "cycles of adaptation" as organic events that "vary in magnitude—both for the technology and the user environment—and elicit different levels of effort and resources. Because this mutual adaptation process is interactive and dynamic, technology may determine structure or vice versa, depending upon when the relationship is observed." Rather than understanding such "disequilibrium between technology and structure" as anything but positive and potentially beneficial, we should view it as being "akin to the creative tension between invention and efficiency" (Leonard-Barton, 1988, p. 252).

While one may expect the term 'adaptation' to carry some degree of valence, Leonard-Barton, unlike others who "have assumed that adjustments in an innovation as it diffuses are undesirable," intends the term to be taken as neutral. If anything, adaptation, since it feeds productivity, can be understood with a positive connotation, allowing for the phenomenon of "'re-invention,' i.e., alteration of the original innovation as users change it to suit their needs or use it in ways unforeseen by developers" to become incorporated into the process of knowledge transfer (p. 253). It should not be surprising to find that "the general tendency for organizational adjustment [is] to lag behind technological change," given the linear progression of invention, development, and implementation. Leonard-Barton suggests that "better performing organizations synchronize the adaptation of administrative policies with the introduction of the technology" (p. 253). Doing so will offset "temporary losses of productivity" associated with the "implementation of new technologies...Just as invention is often triggered by the recognition of performance gaps, so adaptation is precipitated by implementation misalignments—mismatches between the technology and the organization recognized at the time of initial or trial use" (p. 255). Ultimately, Leonard-Barton establishes that "implementation is innovation." Rather than separating technological creation from technological implementation "as if the transfer to operations required merely fulfilling the original charter," we should consider technology and knowledge transfer to be a "continuous, ongoing dedication to the process of change and the conscious management of mutual adaptation" (p. 265).

An Economic View on Knowledge Transfer and Organizational Structure

Considering knowledge as a commodity that profits from transfer among organizations whose primary concerns are scientific research and technology development, Dasgupta and David (1994) take up "economic analysis of the organization of research within the spheres of science and technology [to emphasize] the point that we are dealing with an interrelated system, [comprising] distinct activities that may reinforce and greatly enrich one another, but, furthermore, that it is a system that remains an intricate and rather delicate piece of social and institutional machinery whose constituent elements also may become badly misaligned" (pp. 489-490).

Dasgupta and David seek to remedy the fact that "economics literature addressed specifically to science and its interdependences with technological progress has been quite narrowly focused, and has lacked an overarching conceptual framework to guide empirical studies and public policy discussions in this area." By offering a "new economics of science" that "makes use of insights from the theory of games of incomplete information…in examining the implications of the characteristics of information for allocative efficiency in research activities, on the one hand, with the functionalist analysis of institutional structures, reward systems and behavioral norms of 'open science' communities—associated with [Mertonian] sociology of science," the authors present an analysis of "the gross features of the institutions and norms distinguishing open science from other modes of organizing scientific research, which shows that the collegiate reputation-based reward system functions rather well in satisfying the requirement of social efficiency in increasing the stock of reliable knowledge." Closer examination, however, of "the detailed workings of the system based on the pursuit of priority are found to cause numerous inefficiencies in the allocation of basic and applied science resources, both within given fields and programs and across time" (p. 487).

The authors assert that quantitative benefits from "a basic research advance…will be impeded to the degree to which property rights in such discoveries are intrinsically difficult to establish and defend, and…the organizational norms within which much of such research is conducted (by academic scientists) inhibits effective assertion of individual property rights that can readily be conveyed to other parties, such as business corporations." Ignoring such difficulties will result in a "societal 'underinvestment' in science" (p. 490). Dasgupta and David employ the term 'information' to mean "knowledge reduced and converted into messages that can be easily communicated among decision agents; messages have 'information content' when receipt of them causes some change of state in the recipient, or action. Transformation of knowledge into information is, therefore, a necessary condition for the exchange of knowledge as a commodity" (p. 491).

'Tacit knowledge' "refers to a fact of common perception that we all are often generally aware of certain objects without being focused on them." Tacit knowledge forms "the context which makes focused perception possible, understandable, and productive…science draws crucially upon sets of skills and techniques—the ingredients of 'scientific expertise'—that are acquired experientially, and transferred by demonstration, by personal instruction and by the provision of expert services (advice, consultations, and so forth), rather than being reduced to conscious and codified methods and procedures" (p. 493). Finding organizational structures and vehicles that can facilitate the transformation of tacit knowledge into commodifiable information will "promote knowledge transfers between university-based open science and commercial R&D [in that] there are no economic forces that operate automatically to maintain dynamic efficiency in the interactions of these two (organizational) spheres." Dasgupta and David warn against "ill-considered institutional experiments, which destroy their distinctive features if undertaken on a sufficient scale," for they "may turn out to be very costly in terms of long-term economic performance" (p. 487).

Academic Research Dissemination and Organizational Change

Moed et al. (1985) break new ground in the area of knowledge as an instrument of organizational change by turning their light on academic research production as the subject of investigation, rather than the role of knowledge in innovative product and process development or the impacts of knowledge transfer on social and environmental conditions within an organization. Their study focuses on "the potentialities of quantitative, literature-based (i.e., bibliometric) indicators as tools for university research-policy." Taking two faculty cohorts at the University of Leiden as their data source, Moed et al. analyzed "the research performance of [the] Faculty of Medicine and the Faculty of Mathematics and Natural Sciences...for the period 1970-80." Occurring because of "the necessity for a large-scale project evaluation resulting from a drastic change in the allocation system at the University of Leiden," Moed et al. examined publication and citation data in order to determine which kinds of quantitative data might hold potential for policy change (p. 131).

The first major clarification the authors make is the distinction between "notion quality," a set of differently defined qualitative values, and "bibliometric indicators," which are quantitative measures. The authors "deliberately avoided the use of the notion quality, since it is virtually impossible to operationalize this general concept. Quality may refer to a variety of values. With regard to scientific research, we can distinguish between cognitive quality, methodological quality, and esthetic quality" (pp. 133-134). Bibliographic indictors can be quantified, and it becomes the responsibility of the researcher to frame the data in ways that contribute to the creation of research policy. Moed et al. "focused on indicators based on the number of times publications are cited in the international scientific literature. We argued that citation counts indicate 'impact' rather than quality. Impact is defined as actual influence on surrounding research activities." This definition is significant to the mission of university researchers, as it suggests that "one should not only require that researchers produce results of some scientific quality, but also that they make their results known to colleagues" (p. 147). That is to say, the quality of a particular research article, as judged by its acceptance through peer review, is not only judged by its cognitive, methodological, and/or esthetic qualities, but it is the extension of these qualities into similar and different domains—expressed quantitatively through citation counts—that contributes to the article's notional quality.

Moed et al. make "a distinction between short- and long-term impact. Short-term impact refers to the impact of researchers at the research front up to a few years after publication of their research results. Looking at impact over a long period offers the possibility of relating impact to 'durability'. This long-term influence of research can only be determined after a (very) long time; however, this period is often too long for university science policy, which is concerned with evaluation of recent research" (p. 147). As resource allocation is one result of determining trend impact, trend analysis "as a past performance evaluation over a period of one decade, and a level analysis to determine policy and resource allocation", the authors suggest that "when used properly, this instrument can be a 'monitoring device' for research management and science policy" (p. 148). Ultimately, "the use of bibliometric data for evaluation purposes carries a number of problems, both with respect to data collection and handling, and with respect to the interpretation of bibliometric results. However...[they] enable research policy-makers to ask relevant questions of researchers on their scientific performance, in order to find explanations of the bibliometric results in terms of factors relevant to policy" (p. 131).

CONCLUSION

The creation of knowledge and its adoption and integration into an organization is necessary to obtain advantage from innovation. It is critical to note that implementation of technical innovations is best viewed as a process of mutual adaptation—that is, the re-invention of the technology and the simultaneous adaptation of the organization to obtain a mutual fit (Leonard-Barton, 1988, p. 253). Organizational theorists have worked to address the limitations in earlier economics-based consideration of innovation through the application of organizational theory and other areas of theory. For example, insights from game theory can be applied to examine the implications of the characteristics of information to allocative efficiency in research activities, and with the functionalist analysis of institutional structures, reward systems, and behavioral norms of 'open science' communities (Dasgupta & David, 1994). By analyzing features of the institutions and norms distinguishing open science from other approaches to the organization of scientific research, it was found that the collegiate reputation-based reward system functions well in increasing the stock of reliable knowledge. In summary, there is an interrelated system, comprising distinct activities that may be synergistic with each other; however, due to the intricacies of the social and institutional structure and interrelations, it is possible for the constituent elements to be poorly aligned.

To attempt to counter some of these problems, research on knowledge-based approaches to understanding coordination within the firm demonstrates the implications of the knowledge-based view for hierarchy, the location of decision-making authority, and the boundaries of the firm (Grant, 1996). Knowledge is seen as belonging to individuals, with the firm providing coordinating mechanisms to integrate the specialist knowledge of employees, resulting in knowledge managers gaining insight into the foundations of the organization's capabilities, the principles of organization design, and the horizontal and vertical boundaries of the firm (Grant, 1996). Nonaka and Konno describe four types of *ba*, a space and foundation for knowledge creation, all of which to different degrees allow "the individual [to realize] himself as part of the environment on which his life depends." By participating in *ba*, one can "transcend one's own limited perspective of boundary. This exploration is necessary in order to profit from the 'magic synthesis' of rationality and intuition that produces creativity" (Nonaka & Konno, 1998, p. 41).

From the perspective of organizational theory, neo-institutional theory has been used to develop a framework to understand organizational change, offering insights into change by defining radical (vs. convergent) change and by indicating the contextual dynamics that precipitate the need for organizational adaptation (Greenwood & Hining, 1996). The relation between institutional theory and change can be considered as three elements:

1. A major source of organizational resistance to change derives from the normative embeddedness of an organization within its institutional context.
2. Radical change, and the pace at which change occurs, varies across institutions due to structural differences, especially due to differences in the extent to which institutions are tightly coupled and insulated from ideas practiced in other sectors.
3. The incidence of radical change, and its pace, will vary within sectors due to differences in internal organizational dynamics. Additional insights are offered by taking an alternative approach to considering competencies and capabilities.

The concepts of core capabilities and core rigidities offer a stark contrast with important implications to organizational theory. Core capabilities (Leonard-Barton, 1992) are knowledge sets that distinguish an organization and provide a competitive advantage. It is proposed that these capabilities consist of:

1. Employee knowledge and skills;
2. Technical systems in which the employee knowledge and skills are embedded;
3. Managerial systems that control the processes of knowledge creation and control; and
4. The values and norms associated with the various types of embodied and embedded knowledge, and with the processes of knowledge creation and control.

These capabilities can hamper the firm, in which case management uses one or a combination of the following tactics: (a) abandonment, (b) recidivism, (c) reorientation, and (d) isolation.

While the consideration of knowledge within an organization is important, the transfer of information between organizations is increasingly important. The importance of sectoral differences in motivating partnerships as well as inter-organizational features of technology cooperation are raised (Hagedorn, 1993). This analysis reveals major differences regarding the research orientation of contractual arrangements and organizationally complex alliances. The roles of alliances are considered further by taking a resource-based view of the firm and dynamic capabilities in explaining knowledge transfer and the existence of strategic alliances (Mowery et al., 1996). Alliance activity promotes increased specialization, and the capabilities of partner firms become more divergent in many of the alliances. Alliances are also considered from the perspective of how they contribute to organizational knowledge (Powell et al., 1996). When the knowledge base of an industry is complex and expanding, the sources of innovation will be found in networks of learning if the expertise is widely dispersed; a finding critical to the idea of a network approach to organizational learning, development of alliances, experience with managing inter-firm relationships, and the establishment of portfolios of collaborative activities.

At a more micro level, the role of teams is also important to knowledge development. Participatory teams play an important role in new product development through the sharing of expertise and the dismantling of bureaucratic organizational hierarchies. Olson et al. (1995) find that due to moderating factors there is more than one type of coordinating structure that is likely to be successful in improving product development through delivering more creative new products, cutting development time, and improving new product success in the marketplace. For example, the degree of innovativeness or newness of the product being developed is an important moderator of the impact of different coordination structures on the new product development process and the associated degree of success.

Continuing with the interaction between organizational structure and new product success, the roles of knowledge transfer and the product creation process to explain how concepts of modularity affect both product design and organizational design have been found to be important. By understanding the interactions between component parts, one can specify and standardize the component interfaces, thereby improving a firm's strategic flexibility so it can respond advantageously to changes in the firm's environment. Modular organizational structure is effective in facilitating forms of coordinated self-organizing processes (Sanchez & Mahoney, 1996). By standardizing the interfaces of these nearly decomposable systems in a product design, hierarchical coordination is created without needing tight coupling of organizational structures or the need to exercise authority on a continual basis. Finally, academic research dissemination and organizational change have clear links.

An alternative approach to the interaction of innovation and organizational change is offered by Moed et al. (1985) through the consideration of the generation of knowledge in an academic institution. The study focuses on the ability of knowledge generation to result in changes to policy and organizational change. Moed et al. considered publication and citation data to determine which kinds of quantitative data might hold potential for policy change in an academic institution. The authors assert the importance of distinguishing between "notion quality," a set of differently defined qualitative values, and "bibliometric indicators," which are quantitative measures. The difference in considering notion quality is important to stress, because it is virtually impossible to operationalize this general concept. Having considered knowledge and change in an organization, attention is directed to innovation strategy.

REFERENCES

Cohen, W., & Levinthal, D. (1989). Innovation and learning: The two faces of R&D. *Economics Journal, 99,* 569-596.

Cohen, W., & Levinthal, D. (1990). Absorptive capacity: A new perspective on learning and innovation. *Administrative Science Quarterly, 35,* 128-152.

Daft, R., & Lewin, A. (1993). Where are the theories of the "new" organizational forms? An editorial essay. *Organizational Science, 4*(4), i-vi.

Dasgupta, P., & David, P.A. (1994). Toward a new economics of science. *Research Policy, 23*(5), 487-521.

Grant, R.M. (1996). Toward a knowledge-based theory of the firm. *Strategic Management Journal, 17,* 109-122.

Greenwood, R., & Hining, C.R. (1996). Understanding radical organizational change: Bringing together the old and the new institutionalism. *Academy of Management Review, 21*(4), 1022-1054.

Hagedorn, J. (1993). Understanding the rationale of strategic technology partnering—interorganizational difference modes of cooperation and sectoral differences. *Strategic Management Journal, 14*(2), 371-385.

Leonard-Barton, D. (1988). Implementation as mutual adaptation of technology and organization. *Research Policy, 17*(5), 251-267.

Leonard-Barton, D. (1992). Core capabilities and core rigidities—a paradox in managing new product development. *Strategic Management Journal, 13,* 111-125.

March, J. (1991). Exploration and exploitation in organizational learning. *Organization Science, 2,* 71-78.

Meyer, J., & Rowan, B. (1977). Institutionalized organizations: Formal structure as myth and ceremony. *American Journal of Sociology, 83,* 340-363.

Moed, H.F., Burger, M.B., Frankfort, J.G., & Van Raan, A.F.J. (1985). The use of bbibliometric data for the measurement of university research performance. *Research Policy, 14*(3), 131-149.

Mowery, D.C., Oxley, J.E., & Silverman, B.S. (1996). Strategic alliance and interfirm knowledge transfer. *Strategic Management Journal, 17,* 77-91.

Nonaka, I., & Konno, N. (1998). "The concept of 'ba': Building a foundation for knowledge creation." *California Management Review, 40*(3), 40-54.

Olson, E.M., Walker, O.C., & Ruekert, R.W. (1995). Organizing for effective new product development: The moderating role of product innovativeness. *Journal of Marketing, 59*(1), 48-62.

Powell, W.W., Koput, K.W., & Smith-Doerr, L. (1996). Interorganizational collaboration and the locus of innovation: Networks of learning in biotechnology. *Administrative Science Quarterly, 41*(1), 116-145.

Sanchez, R., & Mahoney, J.T. (1996). Modularity, flexibility, and knowledge management in product and organization design. *Strategic Management Journal, 17,* 63-76.

Scott, W., & Meyers, J. (Eds.). (1994). *Institutional environments and organizations.* Thousand Oaks, CA: Sage.

Spender, J.C. (1989). *Industry recipes: The nature and sources of managerial judgment.* Oxford: Blackwell.

Thompson, J. (1967). *Organizations in action.* New York: McGraw-Hill.

Van de Ven, A. (1986). Central problems in the management of innovation. *Management Science 32*(5), 590-607.

Van de Ven, A., Delbecq, A., & Koenig, R. (1976). Determinants of coordination modes within organizations. *American Sociological Review, 41,* 322-338.

Chapter VII
Organizational Innovation Strategy

INTRODUCTION

There are three dominant themes that run through this chapter on organizational innovation strategy: the rate and nature of change; attitudes, behaviors, and strategic change; and the role of research in organizational strategy. The first section begins with Fry (1982), who examines the interaction between technology and organizational structure in an effort to uncover how this kind of interaction affects how organizations function.

Ettlie, Bridges, and O'Keefe (1984) then look into the food processing industry as an example of organizations that draw clear distinctions between radical and incremental outcomes to support their innovation process model, one which suggests that any organizational innovation process requires a unique implementation strategy and organizational structure that is responsive to the organizational conditions, rather than a more traditional approach, that can be characterized as incremental as opposed to radical change.

Dewar and Dutton (1986) continue the general discussion of rate of change, presenting a study that contrasts the size of firms to their attitudes toward innovation, and finding that investing resources into people as opposed to infrastructure can be a strong facilitator of innovation adoption. Henderson and Clark (1990) also contribute to the discussion regarding the continuum of incremental-to-radical innovation by taking a very close look at the innovative process—from manufacture to end user sale. Ulrich (1995) concludes the first section by drawing from multiple research domains, such as design, software engineering, and operations management, bringing these strands together in an effort to understand how product architecture affects the performance of manufacturing firms.

The second section, with its focus on strategies, attitudes, and behaviors, and their impact on organizational change, begins with Zmud's (1984) discussion of how the values of upper management—and how an organization's members respond to change—affect the organization's use of process innovations. Damanpour (1991) then tests a hypothesis concerning organizational constructs and how they relate to innovation to identify dimensions of innovation that are derived from determinants such as specialization, professionalism, and managerial attitudes toward change. Maidique and Zirger (1984) report on the results and analysis of two surveys regarding industrial innovation in the United States, intended to identify the conditions that contribute to the success of new products, while Tushman and Anderson (1986) find that technological change that contributes to environmental variation is an important factor in how people respond to innovative practices. Kettinger, Teng, and Gusha (1997) examine the methods, techniques, and tools (MTTs) that provide IS professionals with a quantitative basis for business process reengineering (BPR) and places them within an empirical framework. Prahalad and Hamel (1990) conclude this section, making the claim that innovative strategies, to be successful in the 1990s, include de-emphasizing the role of executives as organizational shape-shifters, and identification, growth, and implementation of organizational core competencies that facilitate innovative product development. Research's role in strategic change is the subject of the final section of the chapter, beginning with Henderson and Cockburn's (1994) data, gathered from ten pharmaceutical companies, that help in determining the role of 'competence' in that industry's research. Looking specifically at component and architectural competence, Henderson and Cockburn demonstrate that these two types of competence can explain the nature of variance in research productivity in that industry. Bettis and Hitt (1995) conclude the chapter as they seek to expose technological trends and other factors contributing to the nature of competition in organizations undergoing strategic change, factors such as the rate of technological change and diffusion, and the intensity of the role and importance of knowledge in an increasingly information-based global environment.

RATE AND NATURE OF CHANGE

Fry (1982) presents the results of his review of empirical studies designed to determine "the extent to which the use of different conceptions, levels of analysis, and measures has influenced findings in research on technology-structure relationships" (p. 532). His goal is to "derive a homogeneous body of technology-structure findings that is independent of conception, level of analysis, or operationalization influences" (p. 533). At the time of his writing, Fry assessed the state of the research "examining the relationship between technology and structure and its impact on the functioning of complex organizations," and determined that "there seems to be considerable confusion and overlap concerning the conceptualization of technology and structure," caused particularly by "tendencies to assume homogeneity within categories of variables and to neglect to draw explicitly the line between categories." In addition, studies of technology and structure "have been conducted at the organization, subunit, and individual levels with no attempt to control for possible level effects" (p. 532).

Fry (1982) uses the term 'technology' to describe "the organizational process of transforming inputs into outputs," which "assumes that organizations are open systems and that processes are carried out on at all organizational levels." He finds a variety of ways in which technology is conceptualized, including "technical complexity, technology and operations variability, interdependence, routine or non-routine, and manageability of raw materials [which] does not explicitly state the relationship between technology and interdependence" (pp. 533-38). Structure is defined by Fry to mean "the pattern of events in social systems. Structure is concerned with the arrangement of people, departments, and other subsystems in the organization." Researchers in this area are most concerned with issues of complexity, formalization, and centralization. The agreement of theoretical domains notwithstanding, Fry identifies "the confusion that may result from attempts to compare and generalize findings of technology-structure relationships across three different organizational levels: (1) the whole organization; (2) the work group or subunit; and (3) the individual" (p. 539).

At the organizational level, researchers have not avoided the pitfall of assuming "that the organizations comprising their sample have a single dominant technology. Researchers in this area primarily have used categorical or simple counting procedures to derive an overall organizational technology score. The technical complexity, operating variability, and operations technology concepts are dominant at this level. There is an assumption in this research that work and structural forms across participants and subunits are homogeneous. However, it has been shown that differentiation is a characteristic of complex organizations and that subunits comprising them may be quite diverse" (Fry, 1982, pp. 539-540). This should not obviate the need of researchers to address "discrepancies in findings observed among technology-structure studies [that] may be the result of using either objective or perceptual measures." Analytic variables, also known as perceptual variables, "describe organizations through mathematical operation on some property of each subunit or subunit member." On the other hand, global or objective organizational variables "are not based on information about properties of individual organizational members." Researchers need to be aware that findings "based on objective measures may be biased or nonsensical because the phenomena under study may be misperceived or misrepresented by spokesmen or records and therefore fail to describe adequately the true technological and structural diversity within the organization" (p. 540).

Fry concludes his review with a discussion of analytical variables, which he finds are subject to aggregation bias. "Aggregated data can be interpreted only when there are no level effects on the independent variable." Since "aggregate level correlations may be higher than the same correlations at the individual level when the relation between two variables is systematically different in different units of aggregation," one might be concerned with "the degree to which properties or perceptions of individuals hold true for groups and organizations [comprising] these individuals, and the extent to which one can make inferences from results obtained at one level to higher levels." Moreover, "studies at the individual and subunit levels typically use perceptual measures of technology and structure and may be subject to aggregation problems" (Fry, 1982, p. 541). Ultimately, Fry determines that, with "some exceptions, strong support was found for the existence of technology-structure relationships" (p. 532).

Radical vs. Incremental Outcomes

Ettlie et al. (1984) present a study based on the examination of the food processing industry which promotes "a general model of the innovation process in organizations that is differentiated by radical versus incremental outcomes" (p. 692). Their study tests a model of "organizational innovation process that suggests that the strategy-structure causal sequence is differentiated by radical versus incremental innovation." In other words, they believe that a valid model of organizational innovation process requires a "unique strategy and structure...for radical innovation, especially process adoption, while more traditional strategy and structure arrangements tend to support new product introduction and incremental process adoption." The authors argue that "radical process and packaging adoption are significantly promoted by an aggressive technology policy and the concentration of technical specialists. Incremental process adoption and new product introduction [alternatively] tends to be promoted in large, complex, decentralized organizations that have market dominated growth strategies" (p. 682). Moreover, "traditional structural arrangements might be used for radical change initiation if the general tendencies that occur in these dimensions as a result of increasing size can be delayed, briefly modified, or if the organization can be partitioned structurally for radical vs. incremental innovation." Adopting a radical process requires "centralization of decision-making" and "movement away from complexity toward more organizational generalists," suggesting "that a greater support of top managers in the innovation process is necessary to initiate and sustain radical departures from the past for that organization" (p. 682).

Previous research has established distinctions between different types of innovation attributes, such as administrative as opposed to process, and radical vs. incremental change. Organizations with clear technology strategies, some of which are addressed in this chapter, are often successful because of their ability to predict innovation and successfully integrate these strategies into their policies and tactical procedures. The authors suggest that "unique strategy and structural arrangements are necessary for radical innovation but existing, marketing oriented strategies, and well-known dimensions of organization structure are associated with incremental innovation." Radical technology innovation adoption, as opposed to incremental introduction and adoption, in organizations in part concerns a firm's willingness to incorporate processes that are "a clear, risky departure from existing practice." When a new technology is introduced to a specific unit or "requires both throughput (process) as well as output (production or service) change," the scope of the change, or even the cost of introducing the new technology may be "sufficient to warrant the designation of a rare and radical, as opposed to incremental, innovation" (Ettlie et al., 1984, p. 683).

Ettlie et al. examined "four variable categories [of] radical innovation: technology policy, concentration of technical specialists, the pre-innovation conditions of champion and technology-organization congruence, and radical process and packaging innovation adoption" (p. 683). When technical specialists and expertise is concentrated in a particular unit or firm, incremental innovation is usually the outcome. This inhibits "radical innovation, especially during an organizational crisis, because this type of innovative effort tends to be institutionalized in an 'organic' organization type, but this does not allow for the possibility that experts can be guided by an aggressive policy" (p. 684). Also, when innovation is the outcome of market forces, it tends to unfold incrementally rather than radically. "In well-coordinated organizations there will be some pressure to capitalize on technological developments in the marketplace, so diversification is likely to be a way of maintaining and justifying high R&D expenditures, and diversification is likely to preoccupy the organization with unique new products" (p. 685).

Ettlie et al. (1984) found support for their theory of "innovation process in organizations that is differentiated by radical versus the incremental outcomes," suggesting that "the strategy-structure causal sequence for radical innovation is markedly different from the strategy-structure sequence for incremental innovation (p. 692). However, the authors caution that there are some "notable exceptions to the general model in these data....When the effects of size are controlled...the relative importance of strategy and structure in both parts of the differentiated model is not eroded." Also, "when other variables are controlled, the concentration of specialists does appear to act in both the radical as well as the incremental innovation process" (p. 694).

The Contexts of Radical and Incremental Process Innovations

Like Ettlie et al., Dewar and Dutton are focused on the contexts in which radical or incremental process innovations are best employed in organizations. Acknowledging that "there may be innovations in products, services, social structure or technology, [their] paper concentrates on the adoption of technological innovations involved in a firm's production processes...that incorporate different levels of new knowledge" (Dewar & Dutton, 1986, pp. 1422-1423). Are there "different models...needed to predict the adoption of technical process innovations that contain a high degree of new knowledge (radical innovations) and a low degree of new knowledge (incremental innovations)"? Dewar and Dutton gather data from 40 footwear manufacturers with varying numbers of technical specialists to determine whether either type of innovation adoption, radical or incremental, is affected by the extensive input of specialist knowledge, and how that input affects the adoption of a particular type.

Dewar and Dutton (1986) find that "larger firms are likely to have both more technical specialists and to adopt radical innovations [but] did not find associations between the adoption of either innovation type and decentralized decision making managerial attitudes toward change, and exposure to external information." Several definitions are necessary to understand the scope and nature of Dewar and Dutton's findings. For them, "innovation [is] an idea, practice, or material artifact perceived to be new by the relevant unit of adoption....Radical innovations are fundamental changes that represent revolutionary changes in technology [and] incremental innovations are minor improvements or simple adjustments in current technology" (p. 1422).

Dewar and Dutton describe the "major difference captured by the labels radical and incremental [as being] the degree of novel technological process content embodied in the innovation and hence, the degree of new knowledge embedded in the innovation." As the degree of novelty can be thought of as existing on a continuum, the definition of innovation becomes important to their findings: "An innovation's placement on this continuum depends upon perceptions of those familiar with the degree of departure of the innovation from the state of knowledge prior to its introduction." At this point, the concepts of organizational complexity and centralization of power come into play regarding innovation. Dewar and Dutton (1986) hypothesize that there should be a discernable relationship between an organization's complexity "(the number of different occupational specialties) and the depth of the organization's knowledge resources (the number of technical or engineering personnel)" and their adoption of innovation (p. 1423). Moreover, "the depth of organizational knowledge should also co-vary with the adoption of radical innovations. The greater the number of specialists, the more easily new technical ideas can be understood and procedures developed for implementing them." By concentrating those with knowledge in environments that promote communication, there "exists the context for a greenhouse effect for the development of and support of new ideas [that is] particularly important when these new ideas represent major modifications in the conceptualization of a production process" (p. 1424).

Following this logic, "complexity and knowledge depth should be less important for incremental innovations because adoption of these types [requires] less knowledge resources in the organization for development or support. Instead, adoption of these kinds of changes should be facilitated by mere exposure to the innovation through contact with the external environment." Acknowledging that there will be radical innovative ideas that also exist within this environment, Dewar and Dutton (1986) argue that "few of these ideas will be adopted, however, unless the organization has the internal knowledge resources (complexity and knowledge depth) to interpret and absorb them" (p. 1424). They expect that a firm's exposure to external information will be of greater value to structures that promote "incremental rather than radical innovations, while complexity and depth of knowledge should be more important for radical and not incremental innovation adoption." The difference between a highly decentralized firm as opposed to one in which management is centralized is that the conversion of attitudes toward change into action is more difficult in the former; Dewar and Dutton find that centralization helps "to moderate the relationship between managerial attitudes and the adoption of radical innovations" (p. 1424). Decentralization, in their view, poses a "greater potential for powerful interest groups to dilute proposed change." Firms that exhibit tendencies toward incremental innovation adoption have "less need to mobilize organizational power to overcome potential resistance since these innovations are less costly and have more predictable outcomes" (p. 1424-1425).

While centralization "accelerates the effects of managerial attitudes toward change on radical innovation adoption," there is no evidence that "there is a direct effect of centralization on the adoption of either innovation type." Previous research indicates that in firms exhibiting a decentralized structure, the initiation stage of an innovation in "which lower levels participate in decisions facilitates the circulation of information, [exposes] decision makers to new technological innovations" (see Hage & Aiken, Chapter III). Conversely, in a more centralized firm during the implementation stage of an innovation roll-out, "the adoption process [is facilitated] by reducing conflict and ambiguity." Dewar and Dutton suggest that if both points are accurate, "the effects may cancel each other, leaving centralization with no effect on innovation. Our argument is that centralization will have a direct negative effect on the adoption of incremental innovations. When decentralization gives individuals at lower levels increased power over their work, they will acquire a sense of work ownership and propose changes for improvement.…Centralization, on the other hand, will facilitate radical innovation adoption because more concentrated power may be needed to overcome opposition to these kinds of changes" (Dewar & Dutton, 1986, p. 1425).

Cautioning that "conclusions regarding causality from this research must be tempered because measures of adoption that took place two years prior to measurement of the independent variables were included in the innovation measures," Dewar and Dutton make clear that "confidence in our findings is enhanced by direct comparison with the only other study published that explicitly assesses the determinants of radical versus incremental innovation adoption (Ettlie et al., 1984)." A comparison of results from both studies, one involving footwear manufacturers and the other, food processing concerns, supports the "generalizability" of findings, given that "several key variables were the same across both studies: organizational size, depth of knowledge resources, complexity and decentralization" (Dewar & Dutton, 1986, p. 1430). Managers who are interested in promoting the adoption of technical process innovations should be less concerned "about modifying centralization of decision making, managerial attitudes and exposure to external information [than] managers trying to encourage other types of innovation adoption, e.g., innovations in social services where the factors have been found to be important. Instead, investment in human capital in the form of technical specialists appears to be a major facilitator of technical process innovation adoption" (p. 1422).

Beyond the Radical vs. Incremental Comparison

Henderson and Clark (1990), like several other papers addressed in this chapter, focus their study on the continuum of incremental-to-radical innovation. Their call, however, is to move beyond "models that rely on the simple distinction between radical and incremental innovation, [as they] provide little insight into the reasons why such apparently minor or straightforward innovations should have such consequences." Rather, Henderson and Clark (1990) take "as the unit of analysis a manufactured product sold to an end user and designed, engineered, and manufactured by a single product-development organization" (p. 10). Doing so allows them to show that "the traditional categorization of innovation as either incremental or radical is incomplete and potentially misleading and does not account for the sometimes-disastrous effects on industry incumbents of seemingly minor improvements in technological products." By "distinguishing between the components of a product and the ways they are integrated into the system that is the product 'architecture', [the authors] define them as innovations that change the architecture of a product without changing its components." It is their contention that "architectural innovations destroy the usefulness of the architectural knowledge of established firms, and that since architectural knowledge tends to become embedded in the structure and information-processing procedures of established organizations, this destruction is difficult for firms to recognize and hard to correct." Henderson and Clark's (1990) article discusses how established organizations tackle these "subtle challenges that may have significant competitive implications" (p. 9).

Most research on technical innovation in organizations through the 1980s addressed the "distinction between refining and improving an existing design and introducing a new concept that departs in a significant way from past practice." Henderson and Clark follow Nelson and Winter (see Chapter III), Ettlie et al., Dewar and Dutton, and Tushman and Anderson in defining incremental innovation as the introduction of "relatively minor changes to the existing product, [which exploit] the potential of the established design, and often [reinforce] the dominance of established firms." At the other end of the innovation spectrum is radical innovation, which is "based on a different set of engineering and scientific principles and often opens up whole new markets and potential applications." It has been established by Rothwell et al. (see Chapter III), Tushman and Anderson, and others that although radical innovation "often creates great difficulties for established firms," it also can be "the basis for the successful entry of new firms or even the redefinition of an industry." Generally, incremental innovation "reinforces the capabilities of established organizations, while radical innovation forces them to ask a new set of questions, to draw on new technical and commercial skills, and to employ new problem-solving approaches" (Henderson & Clark, 1990, p. 9).

Henderson and Clark's study of the Xerox copier and the RCA radio receiver identifies "growing evidence that there are numerous technical innovations that involve apparently modest changes to the existing technology but that have quite dramatic competitive consequences." They found that "even after Sony's success was apparent, RCA remained a follower in the market as Sony introduced successive models with improved sound quality and FM capability. The irony of the situation was not lost on the R&D engineers: for many years Sony's radios were produced with technology licensed from RCA, yet RCA had great difficulty matching Sony's product in the marketplace." This is an example of technological innovation in its definitive form: as "innovations that change the way in which the components of a product are linked together, while leaving the core design concepts (and thus the basic knowledge underlying the components) untouched, as 'architectural' innovation" (p. 10).

Viewed as opposite polarities in a defined field, "radical and incremental innovations" are extreme points along the innovation type continuum. "Radical innovation establishes a new dominant design and, hence, a new set of core design concepts embodied in components that are linked together in a new architecture.

Incremental innovation refines and extends an established design" (p. 11). The authors caution that if "a particular innovation is architectural [it] may be screened out by the information filters and communication channels that embody old architectural knowledge" (p. 17). Organizational segmentation and specialization of knowledge, and reliance on "standard operating procedures to design and develop products" depend greatly upon the "dominant design [remaining stable]. Architectural innovation, in contrast, places a premium on exploration in design and the assimilation of new knowledge" (p. 18).

Henderson and Clark see similarities in their work with that of "Abernathy and Clark (1985) [see Chapter III], who have drawn a distinction between innovation that challenges the technical capabilities of an organization and innovation that challenges the organization's knowledge of the market and of customer needs" (Henderson & Clark, 1990, p. 27). The implications of their architectural innovation concepts include not only "a richer characterization of different types of innovation, but they open up new areas in understanding the connections between innovation and organizational capability. The paper suggests, for example, that we need to deepen our understanding of the traditional distinction between innovation that enhances and innovation that destroys competence within the firm, since the essence of architectural innovation is that it both enhances and destroys competence, often in subtle ways." Fundamentally, the effect of an architectural innovation "depends in a direct way on the nature of organizational learning....Given the evolutionary character of development and the prevalence of dominant designs, there appears to be a tendency for active learning among engineers to focus on improvements in performance within a stable product architecture. In this context, learning means learning about components and the core concepts that underlie" them (p. 28).

Modular Architecture

Ulrich (1995) explores the benefits of understanding product architecture, which is "the scheme by which the function of a product is allocated to physical components." Ulrich argues that "the architecture of the product can be a key driver of the performance of the manufacturing firm, that firms have substantial latitude in choosing product architecture, and that the architecture of the product is therefore important in managerial decision-making." Drawing from research in "design theory, software engineering, operations management and management of product development," Ulrich attempts to "synthesize fragments of existing theory and knowledge into a new framework for understanding product architecture, and to use this framework to illuminate, with examples, how the architecture of the product relates to manufacturing firm performance" (p. 419).

Ulrich adds to the definition of product architecture by indicating that a "typology of product architectures...articulates the potential linkages between the architecture of the product and five areas of managerial importance: (1) product change; (2) product variety; (3) component standardization; (4) product performance; and (5) product development management" (p. 419). A comprehensive understanding of product architecture would include familiarity with "(1) the arrangement of functional elements; (2) the mapping from functional elements to physical components; [and] (3) the specification of the interfaces among interacting physical components" (p. 420). In software development, for example, there are occasions when code authors "providc a vocabulary of standard functional elements, while others rely on users to devise their own. Functional elements are sometimes called functional requirements or functives, and the function diagram has been variously called a function structure, a functional description and a schematic description" (p. 420).

Ulrich describes a scale of abstraction in which function structures can be created. "At the most general level, the function structure for a trailer might consist of a single functional element—'expand cargo capacity'. At a more detailed level, the function structure could be specified as consisting of...functional elements [such as] connect to vehicle, protect cargo from weather, minimize air drag, support cargo loads, suspend trailer structure, and transfer loads to road." Once the function structure has been determined, the "second part of the product architecture—the mapping from functional elements to physical components"—comes into play. "A discrete physical product consists of one or more components. For clarity, I define a component as a separable physical part or subassembly. However, for many of the arguments in the paper, a component can be thought of as any distinct region of the product, allowing the inclusion of a software subroutine in the definition of a component. Similarly, distinct regions of an integrated circuit, although not actually separate physical parts, could be thought of as components" (p. 421).

Ulrich draws a distinction in typology between "a modular architecture and an integral architecture. A modular architecture includes a one-to-one mapping from functional elements in the function structure to the physical components of the product, and specifies de-coupled interfaces between components. An integral architecture includes a complex (non one-to-one) mapping from functional elements to physical components and/or coupled interfaces between components" (p. 422).

He also describes two types of product change: "change to a particular artifact over its lifecycle...and change to a product line or model over successive generations" (p. 426). In this environment, standardization "can arise only when: (a) a component implements commonly useful functions; and (b) the interface to the component is identical across more than one different product" (p. 431). Ulrich acknowledges that while "the concept of explicit product architecture is prevalent in large electronic systems design and in software engineering, to my knowledge relatively few manufacturers of mechanical and electromechanical products explicitly consider the architecture of the product and its impact on the overall manufacturing system." His paper is intended to raise the "awareness of the far-reaching implications of the architecture of the product, and [to contribute to] creating a vocabulary for discussing and addressing the decisions and issues that are linked to product architecture [including:] Which variants of the product will be offered in the marketplace? How will the product be decomposed into components and subsystems? How will development tasks be allocated to internal teams and suppliers? What combination of process flexibility and modular product architecture will be used to achieve the desired product variety?" (p. 439).

ATTITUDES, BEHAVIORS, AND STRATEGIC CHANGE

Zmud (1984) brings several related literature strands together—all involving organizational innovation in their different domains—to examine the nature of overlap of their fundamental theories regarding process innovation: "method improvements in task and managerial behaviors, for knowledge work." Zmud's focus is on "the validity of 'push-pull' theory...the importance of top management values...and member receptivity toward change...regarding an organization's use of process innovations" (p. 728).

Given the difficulties involved in bringing new activities and managerial methods into organizations, researchers seek detailed and comprehensive understandings of innovation as it pertains to new product and new process development. Zmud (1984) reviewed the seminal literature of organizational science, research and development management in engineering, and management information systems concerned with "innovation and technology diffusion; however, surprisingly little integration among the three has occurred." Zmud's task in this article is to synthesize central ideas from these domains "in an effort to construct a robust model of innovative behavior." What is the validity of 'push-pull' theory, "that innovation is most likely to occur when a need and a means to resolve that need are simultaneously recognized," and how important is "top management attitude toward an innovation and of organizational receptivity toward change"? To answer these questions, Zmud examined "the diffusion of six modern software practices into 47 software development groups" (p. 727).

Zmud begins with a differentiation between process creation and process diffusion; creation concerns "the development of new methods or machinery," while diffusion is defined as "the adoption of new methods, tools, machinery, etc., to improve work behaviors." Such adoption of innovative processes requires a focus on "product design (and redesign) in an effort to match organizational needs with an emergent technology." At the point when process diffusion is required, "the innovation is usually already well developed when the innovation process commences." Most importantly, innovative practices "often commence when organizational members recognize either a need for change (usually triggered by the emergence of a performance gap, i.e., a problem or opportunity appears) or a new technology (that promises to enhance organizational performance)" (p. 728).

Zmud identifies push-pull theory as being a basic tenet of "engineering/R&D literature"; in fact, it is "a key paradigm for explaining project success or failure" (p. 728). Zmud's findings suggest that "the theory be expanded to include social issues as well as purely technological (performance) concerns." It is 'need-pull' innovation that has "been found to be characterized by higher probabilities for commercial success than have 'technology-push' innovations. While innovation may be induced by either a performance gap or by recognizing a promising new technology, successful innovation is believed to most often occur when a need and the means to resolve it simultaneously emerge." Generally, "the relative importance of particular 'needs' will vary according to the implementation context involved" (p. 728).

'Push-pull' theory failed to be validated by Zmud, who suggests that "top management attitude and organizational receptivity toward change, however, were generally found to influence organizational innovation...significant differences emerged in the factors influencing administrative and technical innovations with organizational receptivity toward change important only for the technical innovations. This suggests that organizational processes facilitating innovation should vary depending on the nature of the innovation involved" (p. 727).

Determinants of Organizational Innovation

Damanpour (1991) conducted a study to test a hypothesis concerning "the relationships between organizational factors and innovation and...the validity of the assumption of instability in the results of innovation research; explore which dimensions of innovation effectively moderate the relationship between innovation and its correlates or determinants; and test some of the existing theories of innovation using the aggregate data" (p. 556). He looked at "the relationships between organizational innovation and 13 of its potential determinants, [which] resulted in statistically significant associations for specialization, functional differentiation, professionalism, centralization, and managerial attitude toward change, technical knowledge resources, administrative intensity, slack resources, and external and internal communication" (p. 555).

Given that an "innovation can be a new product or service, a new production process technology, a new structure or administrative system, or a new plan or program pertaining to organizational members," Damanpour defines innovation "as adoption of an internally generated or purchased device, system, policy, program, process, product, or service that is new to the adopting organization." Because both internal and external environments are subject to constant stress and change, all sorts of organizations "adopt innovations to respond to changes...However, organizational factors may unequally influence innovation in different types of organizations, as extra organizational context and the industry or sector in which an organization is located influence innovativeness" (pp. 556-557).

Damanpour conducted his meta-analysis on the following organizational variables: (1) specialization, (2) functional differentiation, (3) formalization, (4) professionalism, (5) centralization, (6) managerial attitude toward change, (7) managerial tenure, (8) technical knowledge resources, (9) administrative identity, (10) slack resources, (11) external communication, (12) internal communication, and (13) vertical differentiation. His data indicated that "unlike the situation in manufacturing organizations, in service organizations (1) the output is intangible and its consumption immediate, and (2) the producer is close to the customer or client—they must interact for delivery of the service to be complete." The service context is also one in which "technical core employees must deal with client variety and unpredictability, whereas in a manufacturing context, buffering roles reduce uncertainty and disruptions of the technical core." In addition, prior research indicates that "high bureaucratic control (i.e., high formalization and centralization) in turn inhibits innovativeness." Damanpour also correlates these variables to understand whether earlier research's call for "distinguishing types of innovation is necessary for understanding organizations' adoption behavior and identifying [whether] the determinants of innovation in them" is accurate (pp. 558-560).

For Damanpour, "technical innovations pertain to products, services, and production process technology; they are related to basic work activities and can concern either product or process," while "administrative innovations involve organizational structure and administrative processes; they are indirectly related to the basic work activities of an organization and are more directly related to its management." While "managerial attitude toward change and technical knowledge resources have been expected to facilitate radical innovations...structural complexity and decentralization should lead to incremental innovations" (p. 561).

Damanpour also juxtaposes his findings against Burns and Stalker's (1961) notion of mechanistics and organic organizations. In their 1961 study, they "classified organizations according to their patterns of adaptation to technological and commercial change and suggested that mechanistic organizations have lower complexity (hence, lower specialization, differentiation, and professionalism), higher formalization and centralization, lower internal and external communication, and higher vertical differentiation than organic organizations. The key components of the organic organizations Burns and Stalker described are ways of organizing for creativity and innovation" (Damanpour, 1991, p. 579). His findings indicate that "the relations between the determinants and innovation are stable, casting doubt on previous assertions of their instability. Moderator analyses indicated that the type of organization adopting innovations and their scope are more effective moderators of the focal relationships than the type of innovation and the stage of adoption" (p. 555).

Contributing Conditions to New Product Success

Maidique and Zirger (1984) conducted two surveys early in a long-term study of industrial innovation in the United States, as part of the Stanford Innovation Project, in their investigation of innovation success and failure in U.S. electronics firms. The two surveys consisted of "an open-ended survey of 158 new products in the electronics industry, followed by a structured survey of 118 of the original products, both using a pairwise comparison methodology" (p. 192). Of the 60 variables tested in the second survey, 37 were statistically significant and therefore correlated, complicating the results to the degree that there is "no single magical factor that can explain the bulk of our results" (p. 201). There were eight conditions noted that contribute to new production success:

1. "New and growing organizations need an in-depth understanding of...current and potential customers and the general marketplace for its product in order to market 'a product with a high performance-to-cost ratio.'"
2. These same organizations must become adept marketers and commit "a significant amount of [their] resources to selling and promoting the product."
3. The product must provide "a high contribution margin to the firm."
4. The research and development process should be thoroughly planned, and of course, well executed.
5. "The create, make, and market functions are well interfaced and coordinated."
6. The product should be placed into the market early, in relation to its competition.
7. From the perspective of the new product's markets and technologies, the newly developed product will "benefit significantly from the existing strengths of the developing business unit."
8. Overall management support for the product must remain high, "from the development stage through its launch to the market place" (p. 201).

Environmental Variation and Technological Change

Tushman and Anderson (1986) discuss technological change as "a central force in shaping environmental conditions." Technological factors affect organizational forms and technological change affects the size and effectiveness of different populations within organizations. The authors point out that "technological innovation affects not only a given population, but also those populations within technologically interdependent communities." It is this aspect of technological change—how it contributes to environmental variation—that makes it "a critical factor affecting population dynamics" (p. 439). Tushman and Anderson look at patterns of technological change in three diverse industries: "minicomputers," cement, and airlines, to explain how "patterned changes in technology dramatically affect environmental conditions" (p. 440).

Tushman and Anderson take an evolutionary approach, rather than a revolutionary one, to technological change, and echo Kuhn's basic point in the theory of paradigm shifts by pointing out that "technology evolves through periods of incremental change punctuated by technological breakthroughs that either enhance or destroy the competence of firms in an industry." Calling these punctuations "technological discontinuities," the authors discuss how they "significantly increase both environmental uncertainty and munificence" and reach conclusions based on their review of the lifespan of the three industries regarding the relationships between the maturity of a particular industry, and the nature and effects of the discontinuities identified (p. 439).

Adopting Rosenberg's (1972) definition of technology—those "tools, devices, and knowledge that mediate between inputs and outputs (process technology) and/or that create new products or services (product technology)," Tushman and Anderson provide an overview of major discussions of technological change, from Taton (1958) and Schumpeter (1961) who consider "technological change [as] inherently a chance or spontaneous event driven by technological genius," to Gilifillan (1935), where technological change is "a function of historical necessity," to "still others [who] view technological progress as a function of economic demand and growth (Schmookler, 1966; Merton, 1968)" (Tushman & Anderson, 1986, p. 440).

Technological development reifies into what Tushman and Anderson refer to as "dominant design," which "reflects the emergence of product-class standards and ends the period of technological ferment." Dominant design is opposed to "alternative design," which is "largely crowded out of the product class...the dominant design becomes a guidepost for further product or process change." Through the refinement of incremental "technological progress, unlike the initial break-through...interaction of many organizations stimulated by the prospect of economic returns" occurs, resulting in technological development (p. 441).

Essential to Tushman and Anderson's position on technological discontinuities and organizational environments are "two critical characteristics of organizational environments: uncertainty and munificence. Uncertainty refers to the extent to which future states of the environment can be anticipated or accurately predicted...Munificence refers to the extent to which an environment can support growth. Environments with greater munificence impose fewer constraints on organizations than those environments with resource" (p. 445).

Though their data is not conclusive, it is consistent across the industries, that "technological discontinuities exist and that these discontinuities have important effects on environmental conditions" (p. 462). Discontinuities result in "product and process innovation," and "it may be that different kinds of innovation are relatively more important in different product classes. [While] competence-destroying discontinuities are initiated by new firms and are associated with increased environmental turbulence, competence-enhancing discontinuities are initiated by existing firms and are associated with decreased environmental turbulence. These effects decrease over successive discontinuities. Those firms that initiate major technological changes grow more rapidly than other firms" (p. 439). It is Tushman and Anderson's conclusion that "technological change clearly affects organizational environments," and they call for "future research [to] explore the linkage between technological evolution and population phenomena, such as structural evolution, mortality rates, or strategic groups, as well as organizational issues, such as adaptation, succession, and political processes" (p. 463).

Business Process Reengineering

Given the intense interest in business process reengineering (BPR), Kettinger et al. (1997) examine "methods, techniques and tools (MTTs) and places them within an empirical framework" (p. 55). The article is built on the early literature of BPR, including Hammer (1990), who "strongly advocates process 'obliteration,' [calling for] top-down leadership, information technology (IT) enablement, parallel processing, and employee empowerment [see Hammer & Champy, 1993]." Davenport and Stoddard (1994) deem BPR principles such as these as "myths," and Stoddard and Jarvenpaa (1995) "found that Hammer-like 'clean slate' BPR was not typically practiced. They indicate that BPR projects frequently attempt 'revolutionary' (radical) change but because of political organizational and resource constraints, take on 'evolutionary' (incremental) implementations." At the time of their writing, Kettinger et al. (1997) agree with Davenport (1995) that BPR practice "continues to evolve with more emphasis being placed on strategic linkage, smaller projects, fast-cycle methods, and active 'bottom-up' participation" (p. 56).

At this time, "BPR is increasingly recognized as a form of organizational change characterized by strategic transformation of interrelated organizational subsystems producing varied levels of impact." What distinguishes BPR from "past organizational change approaches is its primary focus on the business process....BPR is not a monolithic concept but rather a continuum of approaches to process change." Kettinger et al.'s goal is "to empirically derive a BPR planning framework outlining the stages and activity of a BPR project archetype, [providing] a point of comparison upon which contingent project approaches can be planned" (p. 56).

The authors follow Eisenhardt (1989), which suggests "case- and field-study approaches," conducted a "series of semistructured interviews with BPR experts and vendors...to gain a systematic understanding of BPR MMTs." They gathered "descriptions of 25 BPR methodologies" to derive "a hierarchical MTT map that relates techniques to the BPR project stages and activities, and BPR software tools to techniques." Given the re-engineering consulting firms surveyed, the authors make note that those surveyed "make proprietary BPR methods embodying their own philosophical assumptions, and their consultants tailor their methods to fit clients' unique needs. It was also determined that many of the tools and technology vendors provide BPR services that are based on proprietary methodologies (Kettinger et al., 1997, p. 58).

The research plan consists of seven steps. These include: (1) a literature review of BPR MTTs, (2) collection of "service and product information from MTT consultants and vendors," (3) conducting the semi-structured interviews, (4) creation of "research databases of reengineering MTTs for subsequent analysis," (5) analysis of methodologies and derivation of "a composite BPR project planning framework," (6) examination of the framework to ascertain "reliability and validity," and (7) "map techniques and tools to the S-A [stage-activity] framework." The authors developed a "six-stage, 21-activity, composite S-A framework for BPR"; those six stages include:

1. **Envision:** "A BPR project champion engendering the support of top management...is authorized to target a business process for improvement based on a review of business strategy and IT opportunities in the hope of improving the firm's overall performance."
2. **Initiate:** "This stage encompasses the assignment of a reengineering project team, setting of performance goals, project planning, and stakeholder/employee notification and 'buy-in'."
3. **Diagnosis:** "This stage is classified as the documentation of the current process and sub-processes in terms of process attributes such as activities, resources, communication, roles, IT, and cost. In identifying process requirements and assigning customers value, root causes for problems are surfaced, and non-value-added activities are identified."

4. **Redesign:** At this stage, "a new process design is developed. This is accomplished by devising process design alternatives through brainstorming and creative techniques."
5. **Reconstruct:** "This stage relies heavily on change management techniques to ensure smooth migration to new process responsibilities and human resource roles."
6. **Evaluate:** "This last stage...requires monitoring of the new process to determine if it met its goals and often involves linkages to a firm's total quality programs" (pp. 59, 62).

Regarding the 25 methodologies that were examined and surveyed, most "tend to be strategy driven, with top management interpreting environmental and competitive factors." Most methodologies reviewed "challenge existing assumptions concerning organizational systems...recognize resistance to change and attempt to minimize this through an assessment of cultural readiness and through activities to establish project buy-in...focus on cross-functional and inter-organizational processes [and] take the customer view and leverage IT's coordination and processing capabilities" (p. 62). The authors identify "at least 72 techniques [that] are associated with BPR projects," some of which are "techniques developed in other problem-solving contexts and applying them to BPR." It has been found that "BPR overlaps with socio-technical design (Cherns, 1976) and its later derivations such as soft systems methodology (Checkland, 1981) and promotes the understanding [of] the total work system's technical and social boundaries by employing analysis of social systems boundaries, values, formal and informal information flows, and employees' skill levels....Given the high participation of non-technical personnel on BPR teams, there is a need for more user-friendly and 'media-rich' process capture and simulation packages allowing team members easy visualization and participation in process modeling." Overall, in terms of the tools surveyed, "an expanding suite of tools are being used to provide structure and information management capability in conducting BPR techniques and possess the potential to accelerate BPR projects" (Kettinger et al,. 1997, pp. 62-63).

Different representation techniques are employed in the six stages outlined above. During the 'envision' stage, "search conference...brings stakeholders together to participate in defining both the need for change and how changes should be achieved....The IT/process analysis technique is used to match IT capabilities to a candidate's process requirements....The process prioritization matrix is used in process selection...prepared after top executives establish a firm's critical success factors (CSF) and identify those processes that are essential...or desirable...to achieve the organization's CSFs." The representative technique used in the initiate stage is "quality function deployment (QFD) structures translation of customer needs to process/product characteristics....Team members prioritize a set of customer needs and relate them to process characteristics benchmarked on 'world-class' processes" [see Akao, 1990] (Kettinger et al., 1997, pp. 63,69).

At the diagnosis stage, "process mapping techniques assist project teams in documenting existing processes." During the redesign stage, "creativity techniques such as brainstorming, 'out-of-the-box' thinking, nominal group, and visioning are employed....Process simulation techniques [allow] dynamic modeling to assess process design options. In simulation, process variables such as cycle time, queuing times, inputs/outputs, and resources may be manipulated to provide quantitative analysis of process design scenarios in real-time....Data modeling techniques...utilize the output of process mapping to provide the basis for the data architecture of the new process." Representative techniques used in the reconstruction stage include "force field analysis [which] assists the BPR team in identifying forces resisting the new processes' implementation." Finally, "activity-based costing (ABC) and Pareto diagramming are techniques that allowed reengineering teams to assign activities to cost centers and quantify performance....Pareto diagrams are particularly valuable in graphically ordering problem causes from the most to least significant" (pp. 69-70).

The authors offer several implications of their study, directed toward practice, education, and research. In terms of practice, "planning can be greatly facilitated by developing a customized BPR project S-A methodology and selecting techniques that fit the unique characteristics of the project....IS professionals can make immediate and important contributions to reengineering projects," given their experience with project management, analysis, and design. The techniques suggested through BPR can add to the IS professional's arsenal of skills, thereby gaining "credibility in business planning by capitalizing on their experience in IT/business strategy alignment and their use if IT for competitive advantage." IS education can benefit from the authors' findings, as they provide "a knowledge base of BPR methods [and supply] creative techniques such as force field analysis and nominal group methods" (see Cougar et al., 1993). Moreover, the authors recommend that "IS education should place greater focus on socio-economic systems design and techniques that prepare IS professionals for the softer side of business process change" (Kettinger et al., 1997, pp. 75-76).

As for research implication of their work, "a study is recommended that would further validate the grouping of techniques into common BPR technique classes....A study relating the effectiveness of tool usage to project success would also be interesting. Further research is also recommended in understanding the extent of education and skills improvement needed by BPR practitioners and IS professionals. A final, and probably most important research endeavor is the empirical development of a contingent model predicting project success and the inclusion/exclusion of stages, activities, techniques and tools" (p. 77).

Core Competencies and Innovative Success

Prahalad and Hamel (1990) argue that one of the strategic shifts from the 1980s to the 1990s was away from executive abilities "to restructure, de-clutter, and de-layer their corporations" and toward an "ability to identify, cultivate, and exploit the core competencies that make growth possible." To further their claim, the authors discuss "methods of maximizing the core competencies necessary to compete in a global environment" (p. 79). If "competitiveness derives from an ability to build, at lower cost and more speedily than competitors, [then] the core competencies that spawn unanticipated products" must be developed (p. 81).

To illustrate their point, the authors provide an analogy of a tree to describe the nature of the diversified corporation and its relationship to competitive advantage. "The trunk and major limbs are core products, the smaller branches are business units; the leaves, flowers, and fruit are end products. The root system that provides nourishment, sustenance, and stability is the core competence. You can miss the strength of competitors by looking only at their end products; in the same way you miss the strength of a tree if you look only at its leaves." Considering core competencies to be "the collective learning in the organization, especially how to coordinate diverse production skills and integrate multiple streams of technologies," the authors differentiate between the distribution of knowledge and physical assets: unlike them, "competencies do not deteriorate as they are applied and shared." If this sharing is shaped into a "harmonizing [of] streams of technology...the organization of work and the delivery of value" is maximized. Benefit derives from the sharing of competencies when the skills that constitute them "coalesce around individuals whose efforts are not so narrowly focused that they cannot recognize the opportunities for blending their functional expertise with those of others in new and interesting ways" (p. 82).

Executives can identify core competencies in at least three ways: in terms of "potential access to a wide variety of markets"; in terms of the product or service's contribution "to the perceived customer benefit"; and the degree to which competitors find it difficult to imitate (pp. 83-84). Once identified, core competencies should "constitute the focus for strategy at the corporate level." Product leadership and market share result from successful deployment of "brand-building programs aimed at exploiting economies of scope." When "the company is conceived of as a hierarchy of core competencies, core products, and market-focused business units," it can consider itself strategically positioned for success. In radically decentralized organizations, where "top management [is] just another layer of accounting consolidation," there will be no advance. "Top management must add value by enunciating the strategic architecture that guides the competence acquisition process" (p. 91).

Core Competencies and Economies of Scope

Henderson and Cockburn (1994) investigate "the role of 'competence' in pharmaceutical research" by drawing on "detailed qualitative and quantitative data obtained from 10 major pharmaceutical firms at the program level to show that large firms were at a significant advantage in the management of research through their ability to exploit economies of scope" (p. 64). Their goal is to measure the importance of [different forms of competence] in the context of pharmaceutical research." Their primary distinction of competence types is between 'component' and 'architectural' competence, "and using internal firm data at the program level...show[s] that together the two forms of competence appear to explain a significant fraction of the variance in research productivity across firms" (p. 63).

Henderson and Cockburn build on previously established findings and theories that "have suggested that inimitable firm heterogeneity, or the possession of unique 'competencies' or 'capabilities', may be an important source of enduring strategic advantage," with firm heterogeneity serving as "an important complement to the strategic management field's more recent focus on industry structure as a determinant of competitive advantage" (p. 63). Such heterogenic or "idiosyncratic research capabilities are likely to be a particularly important source of strategically significant 'competence' in science- and technology-driven industries." Additionally, 'architectural competence,' "as captured by our indicators of the firm's ability to integrate knowledge, is positively associated with research productivity" (p. 64).

To turn organizational competence into competitive advantage, it must "be heterogeneously distributed within an industry...be impossible to buy or sell in the available factor markets at less than its true marginal value [and] it must be difficult or costly to replicate." Given that researchers have found "a wide variety of possible sources of heterogeneity fit these criteria, [some] have suggested that unique capabilities in research and development are particularly plausible sources of competitively important competence" (p. 64). The authors define component competence as "the local abilities and knowledge that are fundamental to day-to-day problem solving," with "locally embedded knowledge and skills [being] a source of enduring competitive advantage." Architectural competence is "the ability to use these component competencies—to integrate them effectively and to develop fresh component competencies...make use of [them] in new and flexible ways and to develop new architectural and component competencies as they are required" (pp. 65-66).

Organizational competence includes "architectural knowledge," or "the communication channels, information filters and problem-solving strategies that develop between groups within a problem-solving organization—as well as the other organizational characteristics that structure problem-solving within the firm and that shape the development of new competencies: the control systems and the 'culture' or dominant values of the organization" (p. 66). In that vein, the authors argue "that local capabilities such as proprietary design rules may become so deeply embedded in the knowledge of local groups within the firm that they become strategically important capabilities" (p. 65).

Henderson and Cockburn conclude "that a large proportion of the variance in research productivity across firms could be attributed to firm fixed effects [and] that despite the fact that differences in the structure of the research portfolio had very significant effects on research productivity, variations in portfolio structure across firms were both large and persistent. Both findings are consistent with the existence of exactly the kinds of firm-specific, enduring sources of heterogeneity that are highlighted by the resource-based view of the firm." Moreover, "focusing on 'architectural' or 'integrative' characteristics of organizations can offer valuable insights into the source of enduring differences in firm performance" (p. 64).

Their findings "provide considerable support for the importance of 'competence' as a source of advantage in research productivity, [which] increases with historical success, and to the degree that cumulative success is a reasonable proxy for the kinds of 'local competence' identified in the literature our results suggest that differences in local capabilities may play an important role in shaping enduring differences between firms" (p. 77). Fundamentally, their findings "support the view that the ability to integrate knowledge both across the boundaries of the firm and across disciplines and product areas within the firm is an important source of strategic advantage" (p. 80).

Strategic Management Practices and Discontinuities

Bettis and Hitt (1995) seek to expose "important features of the new competitive landscape and their implications for strategic management practice and research." They do so through discussions of "some of the major technological trends and factors driving strategic change [and] the evolving nature of competition." They focus on "(1) the increasing rate of technological change and diffusion; (2) the information age; (3) increasing knowledge intensity; [and] (4) the emergence of positive feedback industry" (pp. 7-8).

In describing how rapid technological change is altering the "nature of competition in the late twentieth century" as a technological revolution, Bettis and Hitt investigate how "managers and government policy makers face major strategic discontinuities that are" the sources of changes in the scope and nature of competition, and how "rapid development of product and process innovations are becoming increasingly important in many global industries to achieve or sustain a competitive advantage." It is a truism that, "as a result of the increase in speed-based processes...technological changes with strategic implications are occurring at a dizzying pace" (p. 7). Organizations that succeed in "shrinking product development cycles" produce the collateral result of "even shorter product life cycles, concluding in a virtuous (vicious) cycle of continuously faster innovation as a basis for competition" (p. 8).

The shrinking of product life cycles affects the speed with which patented inventions are imitated, with the general result of patents "becoming less effective in protecting new technology. Research has shown that patents are viewed as an effective means of protecting technology only in the chemical and pharmaceutical industries, but are viewed as relatively ineffective in most other industries. Moreover, as a result of the "tendency toward frequent job changes (high job mobility)," the secrets and other technical information that are built into patents "often flow from one U.S. firm to another" (p. 9).

Organizational learning "is a critical component in gaining and/or maintaining competitive advantage in the new technological landscape," and "the strong path dependency associated with technological knowledge creation...means that such current knowledge is a direct function of the firm's formal and informal technological learning in prior time periods" (p. 10). In fact, "firms find it difficult to unlearn past practices, partly because of the self-reinforcing nature of learning." The authors argue that "while it is extremely difficult in the new competitive landscape, firms must develop and exercise the capacity to learn. All firms experience change, but not all learn from it" (p. 15).

The authors consider the new competitive landscape as being compartmentalized into four areas: (1) increasing risk and uncertainty, and decreasing forecastability; (2) the ambiguity of industry; (3) the new managerial mindset; and (4) the new organization and disorganization (p. 11). To this accepted list they add "strategic response capability," or "the generalized ability to respond fast when change or surprise occurs...If firms cannot forecast then they must have the capability to respond quickly" (p. 15). The organizational environment can be described as "highly turbulent and often chaotic [producing] disorder, disequilibrium, and significant uncertainty, 'gusts of creative destruction', as described by Pisano (1990) [see Chapter II], and, thus, discontinuous change [see Tushman & Anderson, 1986, above]. In such an environment, managers must develop new tools, new concepts, new organizations, and new mindsets" (Bettis & Hitt, 1995, pp. 16-17).

CONCLUSION

Organizations are said to need an innovation strategy due to changes in the features of the competitive landscape resulting from an increase in the rate of technological change, diffusion, and knowledge intensity. Bettis and Hitt (1995) note that these changes result in shrinking product development cycles and product life cycles resulting in a cycle of continuously faster innovation now driving competition. As a result of this acceleration to the product life cycle, patents are less effective in protecting new technology,[1] and the trend towards higher mobility increases the flow of technical capabilities and secrets from one firm to another.

To better understand how these changes affect a specific organization, consideration should be given to the relation between technology and structures (Fry, 1982). This is important, because aggregation bias can occur with variables and the relevance of multiple levels of study creates confusion. The term *technology* describes the process of transforming inputs into outputs. *Structure* is defined as a pattern of events in social systems. Structure involves the arrangement of people, departments, and other subsystems in the organization. Devising methodologies and techniques for business process reengineering is the purview of Kettinger et al. (1997). They provide a rationale for employing a variety of technologies and tools that support creative approaches to business process change.

Technology change has a critical environmental effect on both specific organizations and populations of organizations. To better understand this effect, Kuhn's theory of paradigm shifts and punctuated equilibrium theory can be employed. In summary, technology evolves through long periods of incremental change, occasionally punctuated by technological breakthroughs that either enhance or destroy the competence of firms. Tushman and Anderson (1986) refer to these breakthroughs as technological discontinuities. For a firm to take advantage of technological discontinuities, stakeholders must be prepared for significant change at the correct time. Push-pull theory suggests that this sort of change in response to innovation occurs when a need and a means to resolve that need are simultaneously recognized and are supported both by top management attitude and organizational receptivity toward change. Through the consideration of software practices into software development groups, push-pull theory is found to be supported (Zmud, 1984).

Such discussions raise questions about the potential benefits from an innovation strategy and what characteristics within a firm are indicators that organizational innovation is likely to occur. Firstly, the potential benefits of a successful innovation strategy are offered by core competence theory. A core competence can be identified by: (1) potential access to a wide variety of markets, (2) the product or service's contribution to perceived customer benefit, and (3) the degree to which competitors find the competence difficult to imitate (Prahalad & Hamel, 1990).

Because firms that have been successful in adopting a core competence-based strategy have been highly successful generally, the management of a firm should identify and then focus on core competencies.

Having given an example of the benefits possible with an innovation strategy, the determinants of organizational innovation are worth considering. A range of factors were considered through meta-analysis (Damanpour, 1991):

1. Specialization,
2. Functional differentiation,
3. Formalization,
4. Professionalism,
5. Centralization,
6. Managerial attitude toward change,
7. Managerial tenure,
8. Technical knowledge resources,
9. Administrative identity,
10. Slack resources,
11. External communication,
12. Internal communication, and
13. Vertical differentiation.

Of these factors, all but two—formalization and vertical differentiation—were found to have a statistically significant relation with organizational innovation. In other words, 11 of the 13 organizational factors appear to be determinants of organizational innovation.

In addition to structuring an organization so that it adopts and embraces innovation, it is critical to understand which types of innovations offer the most or the greatest potential of benefit. This leads to considerations such as radical, incremental, and architectural. Unique strategy and structural arrangements are necessary for radical innovation, but existing, marketing-oriented strategies and well-known dimensions of organization structure are associated with success in the implementation of incremental innovation (Ettlie et al., 1984). Dewar and Dutton (1986) find that larger firms are likely both to have more technical specialists and to adopt radical innovations. They did not, however, find a relation between the adoption of either innovation type and decentralized decision making managerial attitudes toward change, and exposure to external information. While the terms radical and incremental innovation are useful, it is important not to rely too heavily on the binary measure, since firms often have great difficulty with innovations that can be described as minor improvements. Henderson and Clark (1990) suggest this disconnect can be accounted for by also assessing each innovation from the perspective of an architectural innovation.[2]

Consideration of firms, from a competence perspective, found that *component* and *architectural* competence explain a significant fraction of the variance in research productivity (Henderson & Cockburn, 1994). Furthermore, they found that architectural competence was positively associated with research productivity. Product architecture is the scheme by which the function of a product is allocated to physical components. The architecture of the product is a key driver of the performance of the manufacturing firm, and firms have substantial latitude in choosing product architecture (Ulrich, 1995). Architecture of the product is an important part of managerial decision making. Within architecture there are five areas recognized as having managerial importance:

1. Product change,
2. Product variety,
3. Component standardization,
4. Product performance, and
5. Product development management.

In summary it was found that there are a large number of variables associated with success in innovation (Maidique & Zirger, 1984). Eight conditions that were found to contribute to success are:

1. New and growing organizations need an in-depth understanding of current and potential customers and the general marketplace to market a product with a high performance-to-cost ratio.
2. Organizations must become adept marketers and commit significant resources to selling and promoting the product.
3. Products must provide a high contribution margin.
4. R&D must be thoroughly planned and well executed.
5. Design, operations, and marketing must be well interfaced and coordinated.
6. A product must enter the market early, in relation to its competition.
7. A newly developed product must be synergistic with the existing strengths of its business unit.
8. Overall management support for the product must remain high.

Having considered innovation strategy, the more macro issue of new product development is considered.

REFERENCES

Abernathy, W.J., & Clark, K.B. (1985). Innovation: Mapping the winds of creative destruction. *Research Policy, 14,* 3-22.

Akao, Y. (1990). *Quality function deployment: Integrating customer requirements into product design.* Cambridge, MA: Productivity Press.

Bettis, R.A., & Hitt, M.A. (1995). The new competitive landscape. *Strategic Management Journal,* 16(Summer), 7-19.

Burns, T., & Stalker, G. (1961). *The management of innovation.* London: Tavistock.

Checkland, P. (1981). *Systems thinking, systems practice.* Chichester, UK: John Wiley & Sons.

Cherns, A. (1976). The principles of sociotechnical design. *Human Relations, 28*(8), 783-792.

Damanpour, R. (1991). Organizational innovation: A meta-analysis of effects of determinants and moderators. *Academy of Management Journal, 34*(3), 555-590.

Davenport, T. (1995). Business process reengineering: Where it's been, where it's going. In V. Grover & W.J. Kettinger (Eds.), *Business process change: Concepts, methods and technologies* (pp. 1-13). Hershey, PA: Idea Group.

Davenport, T., & Stoddard, D. (1994). Reengineering: Business change of mythic proportions? *MIS Quarterly, 18*(2), 121-127.

Dewar, R., & Dutton, J.P. (1986). The adoption of radical and incremental innovations—an empirical analysis. *Management Science, 32*(11), 1422-1434.

Eisenhardt, K. (1989). Building theories form case study research. *Academy of Management Review, 14*(4), 532-550.

Ettlie, J.E., Bridges, W., & O'Keefe, R.D. (1984). Organization strategy and structural differences for radical versus incremental innovation. *Management Science, 30*(6).

Fry, L.W. (1982). Technology-structure research: Three critical issues. *Academy of Management Journal, 25*(3), 532-552.

Gilifillan, S. (1935). *Inventing the ship.* Chicago: Follett.

Hammer, M. (1990). Reengineering work: Don't automate, obliterate. *Harvard*

Business Review, 68(4), 104-112.

Hammer, M., & Champy, C. (1993). *Reengineering the corporation: A manifesto for business revolution.* New York: Harper Business.

Henderson, R.M., & Clark, K.B. (1990). Architectural innovation—the reconfiguration of existing product technologies and the failure of established firms. *Administrative Science Quarterly, 35*(1), 9-30.

Henderson, R.M., & Cockburn, I. (1994). Measuring competence—exploring firm effects in pharmaceutical research. *Strategic Management Journal, 15,* 63-84.

Kettinger, W., Teng, J., & Gusha, S. (1997). Business process change: A study of methodologies, techniques and tools. *MIS Quarterly, 21*(1), 55-80.

Maidique, M., & Zirger, B.J. (1984). A study of success and failure in product innovation: The case of the U.S. electronics industry. *IEEE Transactions on Engineering Management, 31*(4), 192-203.

Merton, R. (1968). *Social theory and social structure.* New York: The Free Press.

Nelson, R., & Winter, S. (1977). In search of useful theory of innovation. *Research Policy, 6*(1), 36-76.

Pisano, G. (1990). The R&D boundaries of the firm: An empirical analysis. *Administrative Science Quarterly, 35*(1), 153-176.

Prahalad, C.K., & Hamel, G. (1990). The core competence of the corporation. *Harvard Business Review,* 79-91.

Rosenberg, N. (1972). *Technology and American economic growth.* Armonk, NY: M.E. Sharpe.

Rothwell, R., Freeman, C., Horlsey, A., Jervis, V.T.P., Robertson, A.B., & Townsend, J. (1974). SAPPHO updated—Project SAPPHO phase 2. *Research Policy, 3*(3), 258-291.

Schmookler, J. (1966). *Invention and economic growth.* Cambridge, MA: Harvard University Press.

Schumpeter, J. (1961). *History of economic analysis.* New York: Oxford University Press.

Stoddard, D., & Jarvenpaa, S. (1995). Business process reengineering: Tactics for managing radical change. *Journal of Management Information Systems, 12*(1), 81-108.

Taton, R. (1958). *Reason and chance in scientific discovery.* New York: Philosophical Library.

Tushman, M.L., & Anderson, P. (1986). Technological discontinuities and organizational environments. *Administrative Science Quarterly, 31*(3), 439-465.

Ulrich, K. (1995). The role of product architecture in the manufacturing firm. *Research Policy, 24*(3), 419-440.

Zmud, R.W. (1984). An examination of push-pull theory applied to process innovation in knowledge work. *Management Science, 30*(6), 727-738.

ENDNOTES

1 Except in the chemical and pharmaceutical industries.

2 Architectural innovations destroy the usefulness of the architectural knowledge of established firms; this destruction is difficult for firms to recognize and hard to correct.

Chapter VIII
New Product Development

INTRODUCTION

The articles addressed in this chapter on new product development can be classified in two general categories—papers that address the internal processes that assist or hinder development, and those that focus on factors that contribute to a new product's success or failure in terms of performance and diffusion. We begin with Cooper and Kleinschmidt (1986), who report on the second phase of the New Prod project. Its goal was to examine the nature of the steps that affect the development process and determine how the step-wise structure was modified by the developer companies in order to improve process performance. Clark (1989) looks at project scope, or the extent to which in-house part development affects new product development and overall project performance. The new product development process, as a comprehensive scope of work, is the subject of Millison, Raj, and Wilemon's (1992) discussion, specifically what the tensions and trade-offs are that occur among different functional areas and how they affect innovative product development. Wheelwright and Clark (1992) provide insight into strategies to plan, focus, and control a firm's project development, offering an aggregate project plan that promotes management clearly delineating the roles and steps of each participant's activities. Griffin and Page (1993) offer a practitioner's framework that identifies and coordinates the many measures of product development success and failure, and holds them up against existing measures used by academic researchers. We then move to Souder's (1988) article examining the relationship between R&D groups and marketing groups, the nature of the problems between them, and the structure of potentially effective partnerships.

The second section of the chapter begins with Cooper's 1979 *Journal of Marketing* article, referenced above in the first section. Here, Cooper presents the results of Project New Prod, which was developed to identify the factors that differentiate successful and failing new products. Mahajan and Muller (1979) continue, providing readers with a review of contemporary new product growth models as a basis for understanding recent diffusion models of new product acceptance, helpful to both marketing managers and researchers. Teece then approaches innovation and new product development from an oblique and strategic vantage point in his 1986 *Research Policy* article, suggesting how common and why it is that innovators find themselves in competition with product imitators who benefit more greatly than themselves. For Mansfield (1986), the question is, how do we better understand how and to what extent different industries make use of the patent systems to promote and protect innovation? Cooper and Kleinschmidt (1986) return to this chapter, this time explaining how product superiority is the number one factor influencing commercial success. For them, predevelopment activities of both technical and marketing natures are critical to success in both product development and diffusion. Mahajan and Muller also return to this chapter, with F.M. Bass (1990), providing their insights on the literature on new product diffusion models in marketing. By taking a sociological perspective and grounding their analysis in people's communicative behaviors, these authors unite Bass's innovation diffusion forecasting model with Mahajan and Muller's earlier finding that the objective of a diffusion model is to illustrate the increases in the scope of adopters and predict the nature of the development of an ongoing diffusion process. Montoya-Weiss and Calantone (1994) add to the literature on new product performance by providing a comprehensive overview of research in this area in an effort to identify, determine, and define the factors of new product performance. Brown and Eisenhardt (1995), besides providing one of the more comprehensive literature reviews and analyses of product development, look at empirical studies of product development that focus on the development project as the element of analysis in order to provide a model of factors that contribute to the success of new product development.

INTERNAL PROCESSES

In 1986, Cooper and his collaborator Kleinschmidt reported on the second phase of the New Prod project in their *Journal of Product Innovation Management* article. The goal of the second phase was to look "closely at the new product process: what happens, how well various steps are carried out, and what impact each step has on new product outcomes." To do so, they reviewed "252 new product histories at 123 firms," in which "each company was shown a set of 13 activities which formed a general 'skeleton' of a new product process." Their analysis focuses on "how this structure was modified by the companies and how well various stages of the process were reportedly executed"; their recommendations include that "firms should consider placing more emphasis on market studies, initial screening activities, and preliminary market assessment" (p. 71).

Acknowledging that "product innovation is plagued by high risks: both the large amounts [of capital] at stake and the high probability of failure," Cooper and Kleinschmidt offer the following factors as "fundamental" to new product success:

1. A product differential advantage—"a unique, superior product in the eyes of the customer, a high performance-to-cost ratio, and economic advantages (cost-benefit) to the customer";
2. An understanding of users' needs, wants, and preferences, and a strong market orientation, "with marketing inputs playing an important role in shaping the concept and design of the product";
3. A strong launch effort—selling, promotion, and distribution;
4. Technological strengths and synergy—"a good fit between the product's technology and the technological resources and skills of the firm";
5. Marketing synergy—"a good fit between the marketing, sales force, and distribution needs of the product and the firm's marketing resources and skills";
6. An attractive market for the new product—"a high growth market, a large market, and one with a high long-term potential; a market with weak competition; and one lacking intense competitive activity"; and
7. Top management support and commitment (pp. 71-72).

In the end, however, Cooper and Kleinschmidt suggest that "the overriding finding of the investigation is that new product success is closely linked to what activities are carried out in the new product process, how well they are executed, and the completeness of the process. That is, people—and not solely the nature of the market, the type of technology, or even the synergy or fit between the project and the firm—doing tasks and, most importantly, people doing them well contribute strongly to new product success" (p. 84).

Project Scope

Clark's 1989 *Management Science* paper looks at "one aspect of project strategy, what I shall call project scope: the extent to which a new product is based on unique parts developed in-house." He draws his data from "a larger study of product development in the world auto industry [and looks] at the impact of using off-the-shelf parts and of involving suppliers in development…[finding that] very different structures and relationships exist in Japan, the U.S., and Europe." Clark acknowledges that because "there has been little study of the impact of these differences on development performance…little is known about the effects of different parts strategies (i.e., unique versus common or carryover parts) on development." His analysis provides evidence that there are "differences in strategy, [which] in turn, explain an important part of differences in performance. In particular, I find that a distinctive approach to scope among Japanese firms accounts for a significant fraction of their advantage in lead time and cost" (pp. 1247-1248).

For Clark, "scope has two elements: …The first is the choice of unique versus off-the-shelf parts. Using unique parts adds activities (and cost) to the project and may affect the time required to complete it.…The second element is the choice of supplier involvement" (p. 1248). His analysis "underscores the notion that decisions about the scope of a project have real effects on project performance. The effect of scope on man-hours, for example, is not just an accounting adjustment. Bringing parts engineering in-house and adding work by doing more unique parts design add more engineering hours than one would expect from the amount of the increased workload. The implication is that decisions about scope not only may change the mix of hours (e.g., the ratio of inside to outside hours), but the total engineering effort to develop the product. To the extent that engineering efficiency influences the number and quality of product development projects that a firm undertakes, decisions about scope have strategic significance" (p. 1260).

Milling Down the New Product Development Process

Citing the rapidity with which major manufacturers develop and market new products, Millison et al. (1992) offer insights into the new product development (NPD) process, and discuss "the trade-offs necessary to bring innovative products to rapidly changing markets in the shortest time…and how these functional areas [marketing, research, engineering, and manufacturing] can work together to accelerate NPD (p. 54). The authors' techniques for reducing "the time required to complete the overall NPD cycle [is achieved by] clustering similar tactics": simplify the operation, eliminate delays, eliminate steps, speed up operations, and parallel process (p. 55). Although the article "develops a hierarchy of available NPD acceleration approaches and discusses potential benefits," the authors caution readers about their concomitant "limitations and significant challenges to successful implementation" (p. 53).

While the "typical NPD cycle involves a series of steps such as idea generation, product screening, product development, and commercialization," firms invested in

NPD "must master accelerated product development." The authors cite the literature to support the finding that "the market share advantages that go to 'pioneer' firms [include the] opportunity to create the rules for subsequent competition so that they favor their position" (p. 54). Through a development process geared toward the five NPD cycle acceleration tactics: "(1) simplify, (2) eliminate delays, (3) eliminate steps, (4) speed up operations, and (5) parallel process" (p. 55), managers must balance market demand for quantity with its concomitant demand for quality. "NPD speed that creates quantity without quality should not be the objective of these acceleration approaches. Firms that do not attend to their customers' needs in today's highly competitive environment will not survive. Because most customers are concerned about the quality of their purchases, these acceleration approaches need to be focused on the development of quality products" (p. 64). Speeding up the NPD process "is probably the least effective of our five potential NPD acceleration approaches. This is especially true when related to labor-intensive activities" (p. 65).

Given the considerable time and effort that goes into implementing new product development, "management should not expect to implement these approaches instantly and obtain immediate results" (p. 65). "These approaches may be incorporated alone or in combination and, we emphasize, they may be applied to 'all' of the firm's departments and 'all' of the NPD functions. While many of these techniques are used in manufacturing environments, they have important applications in other functions such as marketing and R&D" (p. 67).

Control and Management Oversight of New Product Development

New product development is also approached by Wheelwright and Clark in their 1992 *Harvard Business Review* article on a strategy to plan, focus, and control a firm's project development. The authors discuss "an aggregate project plan, its component parts, methodology and its value to a firm's product development." Their point is simple but clearly important: "Management must plan how the project set evolves over time, which new projects get added when, and what role each project should play in the overall development effort. The aggregate project plan addresses all of these issues." The aggregate plan approach follows a logical progression of steps, with management keeping a critical and analytic eye on each. "To create a plan, management categorizes projects based on the amount of resources they consume and on how they will contribute to the company's product line. Then, by mapping the project types, management can see where gaps exist in the development strategy and make more informed decisions about what types of projects to add and when to add them" (p. 72).

The authors suggest using two strategies for classifying activities: "the degree of change in the product and the degree of change in the manufacturing process" (p. 73). Additionally, managers should periodically evaluate "the product mix [to keep] development activities on the right track. Companies must decide how to sequence projects over time, how the set of projects should evolve with the business strategy, and how to build development capabilities through such projects" (p. 79).

Generally, "the greatest value of an aggregate project plan over the long term is its ability to shape and build development capabilities, both individual and organizational. It provides a vehicle for training development engineers, marketers, and manufacturing people in the different skill sets needed by the company" (p. 81). Given the difficulties involved in establishing and implementing an aggregate project plan, "working through the process is a crucial part of creating a sustainable development strategy." Managers should keep in mind that most likely the "plan will change as events unfold and managers make adjustments. But choosing the mix, determining the number of projects the resources can support, defining the sequence, and picking the right projects raise crucial questions about how product and process development ought to be linked to the company's competitive opportunities. Creating an aggregate project plan gives direction and clarity to the overall development effort and helps lay the foundation for outstanding performance" (p. 82).

Measuring New Product Development Success and Failure

Griffin and Page (1993) present "an interim report that identifies and structures into a useful framework the myriad measures of product development success and failure [S/F], and compares those measures used by product development practitioners and managers to those used by academic researchers. The PDMA [Product Development and Management Association] task force hopes that ultimately this research on how S/F is measured will bring academic researchers and industry practitioners together onto a common ground for evaluating S/F" (p. 292).

Griffin and Page "sought to identify all currently used measures, organize them into categories of similar measures that perform roughly the same function, and contrast the measures used by academics and companies to evaluate new product development performance. The authors compared the measures used in over 75 published studies of new product development to those surveyed companies say they use." One of the authors' findings is that, while "firms generally use about four measures from two different categories in determining product development success...academics and managers tend to focus on rather different sets of product development success failure measures. Academics tend to investigate product development performance at the firm level, whereas managers currently measure, and indicate that they want to understand more completely, individual product success" (p. 291).

To sum up their findings and conclusions, Griffin and Page "have found the following:

- Measuring S/F generally is multidimensional.
- Five independent dimensions of S/F performance have been identified: firm-, program-, and product-level measures, and measures of financial performance and customer acceptance.
- Practitioners use about four measures from a total of two different dimensions, most frequently customer acceptance and financial permanence.
- Researchers use slightly fewer measures, about three, from one to two dimensions. The particular dimensions used differ across three different clusters of researcher focus.
- Researchers have focused more on overall firm impacts of S/F, whereas companies focus on the S/F of individual projects. Ultimately, our goal is to be able to recommend what categories of product development success and failure should be measured, and which measures within categories are the most powerful indicators of S/F" (p. 305).

Cooperation Between R&D and Marketing

Souder's 1998 article discusses the researcher's findings after review of a database of 189 new product development innovation projects in which managers reported "incidences of different types of problems" between R&D groups and marketing groups that affected project outcomes. Souder uses these findings to develop "a number of recommendations for increasing the success rates of innovation projects by using a model that improves conditions at the R&D/marketing interface" (p. 6). The intention is to provide managers with a process whereby effective partnerships between R&D and marketing groups can achieve "a more collaborative, partnership role" (p. 13).

The eight guidelines are as follows:

1. *Break Large Projects into Smaller Ones.* Souder found that more than 75% of projects that had nine or more people assigned to a particular project task had "interface" problems, while five or fewer project members rarely experienced problems when R&D and marketing staff were teamed together. "The smaller number of individuals and organizational layers on the small projects permitted increased face-to-face contacts, increased empathies and easier coordination."

2. *Take a Proactive Stance Toward Interface Problems.* In other words, do not sit back and let incipient, inchoate interpersonal problems fester. Interface problems are avoidable when team members and managers are "aggressively seeking out and facing such problems head-on...[when they] openly criticized and examined their behaviors."
3. *Eliminate Mild Problems Before They Grow into Severe Problems.* Much like point 2 above, once a team recognizes that there are "mild," or not fully formed and articulated problems in terms of interpersonal relationships, it is best to tackle them and eliminate them before they grow into intractable differences between group members.
4. *Involve Both Parties Early in the Life of the Project.* Earlier research indicates that early involvement of both R&D and marketing personnel in the decision-making process leads to project completion rather than stagnation and failure, and it curtails instances of "lack of appreciation and distrust" among project staff.
5. *Promote and Maintain Dyadic Relationships.* Souder defines "dyad" as "a very powerful symbiotic, interpersonal alliance between two individuals who become intensely committed to each other and to the joint pursuit of a new product idea. Dyads are observed any time persons with complementary skills and personalities are assigned to work together and given significant autonomy." He suggests that dyads "encourage innovation" as well as offer an example of effective team dynamics for a potentially "much wider circle of interrelationships between R&D and marketing" personnel.
6. *Make Open Communication an Explicit Responsibility of Everyone.* An environment in which project team members are encouraged to voice their ideas and suggest remedies to problems as they arise is one that leads to successful delivery of innovative products. Souder offers an example of one company's "Open Door" policy, in which "a history of poor R&D/marketing interfaces" was reversed with the implementation of "quarterly information meetings between R&D and marketing, daylong and weeklong exchanges of personnel, periodic gripe sessions, and the constant encouragement of personnel to visit their counterparts. Every employee was formally charged with the responsibility of playing a role in this Open Door policy. Moreover, each employee's success in meeting this responsibility was formally evaluated at the end of each quarter." Although there was reluctance on the part of many to engage the new policy, "the examples set by a few diligent individuals eventually spread."
7. *Use Interlocking Task Forces.* By "interlocking," Souder means the presence of highly place officers and hands-on project management personnel working as a steering committee for real-time project-critical tasks and processes. As

the project grows and takes on different forms, the "marketing and R&D task force memberships [change] as the project [metamorphoses] over its life cycle." The idea is for each individual, with a particular set of strengths, expertise, and responsibilities, to reinforce the efforts of the others on the task force, and all task force members participate at the specific times when their skill sets can have the most positive impact on "R&D/marketing harmony and new product development success."

8. *Clarify the Decision Authorities.* Decision authority is an agreed-upon set of guidelines and responsibilities that R&D and marketing team members can use to steer the innovative product development process. When all team members know what their purviews are and who is ultimately responsible for decisions of specific sorts, it "can contribute enormously to clarifying the roles between R&D and marketing [and foster] a sound foundation for the avoidance of many time-consuming conflicts" (pp. 13-14).

Souder concludes "that R&D and marketing managers should jointly work together to help avoid disharmonies in seven ways": by establishing the fact that the interface problems discussed above "naturally occur"; by encouraging their personnel to "be sensitive [and aware] to the emergence of R&D/marketing interface problems"; "to give equal credit and public praise to their R&D and marketing personnel in order to eliminate jealousies that might form a basis for severe disharmony"; to encourage in tangible and obvious ways collaboration between R&D and marketing; to take any opportunity to use cross-functional teams; to be proactive and direct in efforts to eradicate mild problems when they first surface; and "managers must also be aware that there is such a thing as too much harmony: R&D and marketing personnel can become too complacent with each other" (p. 18).

NEW PRODUCT DIFFUSION: SUCCESS AND FAILURE

In Cooper's 1979 *Journal of Marketing* article, he presents the results of a study, titled New Prod, "whose purpose was to identify the major factors which differentiate between successful and unsuccessful new industrial products" (p. 93). His contribution to the literature is his "identifying 18 underlying dimensions that capture much of the new product situation...dimensions that can be used to characterize and perhaps cluster new product projects. (pp. 100, 102). Cooper uses multivariate methods "to probe this success/failure question," identifies the "dimensions underlying success and failure," and indicates what the "dominant role of product strategy" is and why there's a "need for a strong market orientation" (p. 93).

At the top of the list of "dimension[s] leading to new product success is Product Uniqueness and Superiority," followed by "Market Knowledge and Marketing Proficiency [as playing] a critical role in new product outcomes" (p. 100). Taking third place is "Technical and Production Synergy and Proficiency. Projects where such synergy and proficiency existed were undertaken in firms with a particularly strong and compatible technical engineering and production and resource base" (p. 101). Dimensions contributing to new product failure include: "having a high-priced product, relative to competition (with no economic advantage to the customer); being in a dynamic market (with many new product introductions); [and] being in a competitive market, where customers are already well satisfied." Also contributing to new product success are "Marketing and Managerial Synergy; Strength of Marketing Communications and Launch Effort; [and] Market Need, Growth, and Size" (p. 101).

Diffusion Models of New Product Acceptance

Mahajan and Muller's 1979 article provides readers with a review of contemporary new product growth models in hopes of offering "marketing managers and researchers...a simple and systematic overview of the development of diffusion models of new product acceptance" (p. 55). They identify "the underlying assumptions of these models...[where] these models [have] been applied...[how] the proposed models [are] different from each other...[what] the shortcomings of these models [are] and...what directions need to be followed to make these models theoretically more sound and practically more effective and realistic" (p. 55).

Mahajan and Muller base their views in the presumption that "the first-purchase diffusion models of new product acceptance" assume that, in the product-planning "horizon being considered, there are no repeat buyers and purchase volume per buyer is one unit." They concern themselves with "the total flow of customers and the rate of flow of customers, across three distinct segments of the market: the untapped market, the potential market and the current market" (p. 56). The "basic diffusion models consider only two segments in the diffusion process" when examining "the growth of the first-time buyers of a product: the potential market and the current market; two transfer mechanisms...influence the potential customers to adopt the product: mass-media communication [and] word-of-mouth communication" (p. 57). The authors note that "these models assume a constant total population of potential customers over the entire life of the product" (p. 58).

The authors conclude that the "basic diffusion models of new product acceptance...are of little use to the new product manager since they consider diffusion as a function of time only and the marketing program of a company does not enter explicitly as a variable inhibiting the evaluation of the effect of different marketing strategies on the product growth" (p. 60).

Innovators, Imitators, and Successful New Product Development

Teece approaches innovation and new product development from an oblique and strategic vantage point in his 1986 *Research Policy* article. Pointing out how common it is for innovators, whom he defines as "those firms that are first to commercialize a new product or process in the market" only to find that "competitors/imitators have profited more from the innovation than the firm first to commercialize it," Teece attempts to "explain why a fast second or even a slow third might outperform the innovator," by pointing out how and why it is not always the case that "developing new products, which meet customer needs, will ensure fabulous success. It may possibly do so for the product, but not for the innovator." What are the factors that "determine who wins from innovation: the firm which is first to market, follower firms, or firms that have related capabilities that the innovator needs"? (p. 285). The short answer for Teece is: "Business strategy—particularly as it relates to the firm's decision to integrate and collaborate...when imitation is easy, markets don't work well, and the profits from innovation may accrue to the owners of certain complementary assets, rather than to the developers of the intellectual property." While some firms heed the need "to establish a prior position in these complementary assets...innovators with new products and processes which provide value to consumers may sometimes be so ill positioned in the market that they necessarily will fail....Innovating firms without the requisite manufacturing and related capacities may die, even though they are the best at innovation" (p. 285).

Teece identifies "three fundamental building blocks [that] must first be put in place: the appropriability regime, complementary assets, the dominant design paradigm" (p. 286). The first block, a regime of appropriability, "refers to the environmental factors, excluding firm and market structure, that govern an innovator's ability to capture the profits generated by an innovation." Dominant paradigms arrive in periods of "scientific maturity," and it is then that, with "agreed upon 'standards' by which what has been referred to as 'normal' scientific research can proceed" (p. 287).

Even though prior research has "provided a treatment of the technological evolution of an industry" which appears to parallel "legal instruments [and the] nature of technology," such a framework "does not characterize all industries. It seems more suited to mass markets where consumer tastes are relatively homogeneous. It would appear to be less characteristic of small niche markets where the absence of scale and learning economies attaches much less of a penalty to multiple designs." With maturity, however, "and after considerable trial and error in the marketplace, one design or a narrow class of designs begins to emerge as the more promising. Such a design must be able to meet a whole set of user needs in a relatively complete fashion." In the natural course of product evolution, "once the product design stabilizes, there is likely to be a surge of process innovation as producers attempt to lower production costs for the new product." Teece's primary point, however, is that "the successful commercialization of an innovation requires that the know-how in question be utilized in conjunction with other capabilities or assets" (pp. 287-288).

Teece distinguishes between "generic assets [which] are general-purpose assets [that] do not need to be tailored to the innovation in question" and "specialized assets...where there is unilateral dependence between the innovation and the complementary asset. Co-specialized assets are those for which there is a bilateral dependence" (p. 289). One sure way for an innovator to be "almost assured of translating its innovation into market value for some period of time" is to hold a patent or copyright protection (p. 290). Innovators who wish to retain control over their new products and processes must take control over "access to complementary assets, such as manufacturing and distribution," or their interests will be overtaken by "imitators, and/or...the owners of the complementary assets that are specialized or co-specialized to the innovation" (p. 292).

Teece's strategy is not limited to any size firm or niche markets. "[Many] small entrepreneurial firms which generate new, commercially valuable technology fail while large multinational firms, often with a less meritorious record with respect to innovation, survive and prosper. One set of reasons for this phenomenon is now clear. Large firms are more likely to possess the relevant specialized and co-specialized assets within their boundaries at the time of new product introduction." "In industries where legal methods of protection are effective, or where new products are just hard to copy, the strategic necessity for innovating firms to integrate into co-specialized assets would appear to be less compelling than in industries where legal protection is weak. In cases where legal protection is weak or nonexistent, the control of co-specialized assets will be needed for long-run survival" (p. 301). Ultimately, "ownership of complementary assets, particularly when they are specialized and/or co-specialized, help establish who wins and who loses from innovation. Imitators can often outperform innovators if they are better positioned with respect to critical complementary assets. Hence, public policy aimed at promoting innovation must focus not only on R&D, but also on complementary assets, as well as the underlying infrastructure" (p. 304).

New Product Development and the Benefits of Patents

Mansfield (1986) begins by asking, rhetorically, the percentage of "patentable inventions that are patented." With no existing data available to address this question, Mansfield refines the goal of his research: "To achieve a better understanding of the extent to which various industries and types of firms make use of the patent system" (p. 176), and therefore the research questions, as well: "To what extent would the rate of development and introduction of inventions decline in the absence of patent protection? To what extent do firms make use of the patent system, and what differences exist among firms and industries and over time in the propensity to patent?" The paper provides a summary of Mansfield's "empirical investigation based on data obtained from a random sample of 100 U.S. manufacturing firms" (p. 173).

Conventional wisdom has it that "patent protection tends to be more important to smaller firms than to larger ones, [but] the existing evidence on this score is weak and sometimes contradictory" (p. 175). In fact, some firms, instead of seeking patents, "rely instead on trade secrets, because technology is progressing so rapidly that it may be obsolete before a patent issues, because it is very difficult to police the relevant subject matter, or for other reasons" (p. 176). Although trade secrets may prove to be a less expensive method of protecting new, inventive ideas and technologies, maintaining trade secrets does not provide the actionable protection of a patent; the question for Mansfield becomes: What industries tend to seek patents and which "have become more disillusioned with the patent system, [devising] other ways of protecting their technology that are more cost-effective than patents?" (p. 178).

Leading the list of 12 industries, whose firms comprised the random 100 analyzed, were the pharmaceutical and chemical fields, in which "patent protection was judged to be essential for the development or introduction of 30% or more of the inventions." Following these were "petroleum, machinery, and fabricated metal products, [where] patent protection was estimated to be essential for the development and introduction of about 10-20% of their inventions." For the "remaining seven industries (electrical equipment, office equipment, motor vehicles, instruments, primary metals, rubber, and textiles), patent protection was estimated to be of much more limited importance in this regard" (p. 174).

Given that patents are "defended at least partly on the grounds that [they increase] the rate of innovation, the present study indicates that [their] effects in this regard are very small in most of the industries we studied." Trade secret protection is less preferred when patents are a viable option, "but...despite the frequent assertions that firms are making less use of the patent system than in the past, the evidence does not seem to bear this out" (p. 180).

Linking Project Outcomes to Descriptors

Cooper and Kleinschmidt return again in 1987 to the pages of the *Journal of Product Innovation Management,* this time to discuss findings resulting from their efforts to "resolve the success-vs.-failure issue by studying 203 new products—both winners and losers—that were launched into the marketplace." A thorough review of past studies and literature resulted in a conceptual model of new product success and failure—a model that links project outcomes to a number of key project descriptors (p. 169).

Their foray into the determinants of new product success and failure results in the conclusion that "product superiority is the number one factor influencing commercial success and that project definition and early, predevelopment activities are the most critical steps in the new products development process. Success, they argue, is earned. "It is not the ad hoc result of situational or environmental influences. Synergy, both marketing and technical, is crucial" (p. 169).

Characteristics identified as "the most important discriminators...between commercial successes and failures" were:

1. Understanding of users' needs,
2. Attention to marketing and publicity,
3. Efficiency of development,
4. Effective use of outside technology and external scientific communication, and
5. Seniority and authority of responsible managers (p. 171).

Identifying these determinants of success "has also lent support to a model of new product outcomes, and yielded results that have action implications to managers" (p. 183).

Refining New Product Diffusion Models

Mahajan and Muller return to the *Journal of Marketing* in 1990, this time with F.M. Bass, to provide a "critical evaluation" of the literature on new product diffusion models in marketing since the earlier Mahajan and Muller article in 1979 and Bass' input into this domain in 1969. Their assessment of the work done in this area over the course of the late 20th century demands, from their perspective, the creation of "a research agenda to make diffusion models theoretically more sound and practically more effective and realistic" (Mahajan, Muller, & Bass, 1990, p. 1).

The authors take a sociological perspective by grounding their analysis in people's communicative behaviors, finding that "members of a social system have different propensities for relying on mass media or interpersonal channels when seeking information about an innovation. Interpersonal communications, including nonverbal observations, are important influences in determining the speed and shape of the diffusion process in a social system." The fact that "innovation diffusion theory has sparked considerable research among consumer behavior, marketing management, and management and marketing science scholars" suggests that even though researchers in different disciplines have different goals, there is a need for a common base from which to examine a theory's constructs. "Researchers in consumer behavior have been concerned with evaluating the applicability of hypotheses developed in the general diffusion area to consumer research...The marketing management literature has focused on the implications of these hypotheses for targeting new product prospects and for developing marketing strategies aimed at potential adopters" (p. 1).

Noting the ubiquity of the Bass innovation diffusion forecasting model—a formula to determine the number of users of a particular technology at a given time that employs coefficients of external and internal influence along with market potential—found in research on "retail service, industrial technology, agricultural, educational, pharmaceutical, and consumer durable goods markets," the authors incorporate "Mahajan and Muller's (1979) [statement] that the objective of a diffusion model is to present the level of spread of an innovation among a given set of prospective adopters over time. The purpose of the diffusion model is to depict the successive increases in the number of adopters and predict the continued development of a diffusion process already in progress" (Mahajan et al., 1990, p. 2).

For the authors, the "basic structure of the Bass diffusion model [raises] three questions:

- How does the Bass model compare with the classical normal distribution model proposed by Rogers (1983)?
- Is the Bass model complete in capturing the communication structure between the two assumed distinct groups of innovators and imitators?
- How can the Bass model, which captures diffusion at the aggregate level, be linked to the adoption decisions at the individual level?"

Rogers relies on "two basic statistical parameters of the normal distribution—mean and standard deviation—[to propose] an adopter categorization scheme dividing adopters into...categories of Innovators, Early Adopters, Early Majority, Late Majority, and Laggards" (Mahajan et al., 1990, p. 5).

Both the Rogers and Bass models posit that "at any time *t*, a potential adopter's utility for an innovation is based on his uncertain perception of the innovation's performance, value, or benefits." Mahajan et al. point out that the "potential adopter's uncertain perceptions about the innovation, however, change over time as he learns more about the innovation from external sources (e.g., advertising) or internal sources (e.g., word of mouth). Therefore, because of this learning, whenever his utility for the innovation becomes greater than the status quo (he is better off with the innovation), he adopts the innovation" (p. 6).

Even though "innovation diffusion models traditionally have been used in the context of sales forecasting[,] sales forecasting is only one of the objectives...perhaps the most useful applications of diffusion models are for descriptive and normative purposes. Because diffusion models are an analytical approach to describing the spread of a diffusion phenomenon, they can be used in an explanatory mode to test specific diffusion-based hypotheses" (p. 15).

Determining New Product Performance

Montoya-Weiss and Calantone (1994) add to the literature on new product performance by providing a comprehensive overview of research in this area in an effort to identify "the nature of research on the determinants of new product performance and define the set of factors identified in this review" (p. 398). Their specific contribution, as the editors of the *Journal of Product Innovation Management* describe it, is that they have "developed quantitative comparisons of [their] results, which, although cumbersome, provide a look at the persistent exploratory nature of this research. They report a wide variation in results that are surprisingly non-convergent" (p. 397).

Montoya-Weiss and Calantone find that, generally, "new product success studies identify characteristics and factors leading to success, whereas failure studies provide retrospective analyses of past failures in an effort to identify the determinants of failure or common pitfalls and problems in the development process," suggesting that success studies provide less detail and explanation as to why and how new products are successful in comparison to analyses of failure (p. 398). There is a great amount of diversity and considerable inconsistency regarding contributing factors to new product performance in the literature, including studies that focus on "research methodology, data set characteristics, and variable operationalizations" (p. 411).

Montoya-Weiss and Calantone do come to "several general conclusions...regarding the content and nature of empirical research on new product performance...First, although there is some consistency as to which factors are considered by researchers, the range of factors included in the typical set is indeed narrow...A second result of this review is the discovery that some factors have not been studied extensively enough to draw strong conclusions regarding their impact on performance...Third, our findings suggest that the effects of certain strategic and development process factors upon new product performance have been most strongly linked to performance in the literature....A fourth outcome of this review is a clear demonstration of a need for more correlational analyses and tests of differences between success/failure groups" (p. 412).

A Comprehensive View of New Product Development

Brown and Eisenhardt's 1995 *Academy of Management Review* article presents one of the more comprehensive literature reviews and analyses of product development. With a "focus on normative empirical studies of product development in which the development project is the unit of analysis," the authors "synthesize [their] research findings into a model of factors affecting the success of product development [and] indicate potential paths for future research" that are "centered on the effects that the development process and product concept have on product success, patterns of organizing product-development work, strategic management, and customer/supplier involvement" (pp. 344-345). With a substantial body of research in the "organizations-oriented tradition, [which] focuses at a micro-level regarding how specific new products are developed," Brown and Eisenhardt concentrate on "the structures and processes by which individuals create products. "In this article, we focus on this area of the broader innovation literature" (p. 343).

It is generally accepted that innovation research falls into two camps. The "economics-oriented tradition examines differences in the patterns of innovation across countries and industrial sectors, the evolution of particular technologies over time, and intra-sector differences in the propensity of firms to innovate...[The] organizations-oriented tradition focuses at a micro-level regarding how specific new products are developed" (p. 343). Another fundamental concept regarding product development is its importance, "probably more than acquisition and merger, [as] a critical means by which members of organizations diversify, adapt, and even reinvent their firms to match evolving market and technical conditions...Thus, product development is among the essential processes for success, survival, and renewal of organizations, particularly for firms in either fast-paced or competitive markets" (p. 344).

The authors organize product development literature into "three streams of research: product development as rational plan, communication web, and disciplined problem solving." After synthesizing their "research findings into a model of factors affecting the success of product development" in order to highlight "the distinction between process performance and product effectiveness and the importance of agents, including team members, project leaders, senior management, customers, and suppliers, whose behavior affects these outcomes, [they] indicate potential paths for future research based on the concepts and links that are missing or not well defined in the model" (p. 343). Acknowledging that "there are overlaps in focus across the streams (e.g., all streams investigate how different players, processes, and structures affect performance), research within each stream centers on particular aspects of product development." Differences among the streams include rational plan research focusing on "a very broad range of determinants of financial performance of the product, whereas the communication web work concerns the narrow effects of communication on project performance. Disciplined problem solving centers on the effects of product—a development team, its suppliers, and leaders on the actual product-development process" (p. 345). The primary benefit of the rational plan perspective arises out of its inherent "exploratory and a theoretical" nature, which "helps to broadly define the relevant factors for product-development research. The communication web stream complements this theoretical view by relying on information-processing and resource dependence theoretical perspectives in the context of traditional research studies. The disciplined problem-solving stream takes the theoretical perspective of information processing one step further to problem-solving strategies, using a progression from inductive to deductive research and an emphasis on global industry studies" (pp. 345, 347-348).

The rational plan perspective "emphasizes successful product development which is the result of (a) careful planning of a superior product for an attractive market and (b) the execution of that plan by a competent and well-coordinated cross-functional team that operates with (c) the blessings of senior management." Researchers invoking this stream are intent on "discovering which of many independent variables are correlated with the financial success of a product-development project" (p. 348). Of principal importance for product success "was product advantage…intrinsic value of the product, including unique benefits to customers, high quality, attractive cost, and innovative features, was the critical success factor. Such products were seen as superior to competing products and solved problems that customers faced." Also important "was predevelopment planning. This included developing a well-defined target market, product specifications, clear product concept, and extensive preliminary market and technical assessments. Other internal organization factors also were important, including cross-functional skills and their synergies with existing firm competencies. Top management support also was important, but less so than these other factors. Finally, market conditions also affected product success" (p. 351). Brown and Eisenhardt found, surprisingly, that "market competitiveness had no relationship with product success." Speed of product development, on the other hand, was positively correlated with success. For the rational plan stream of research, rational planning and execution results in successful product development. "That is, successful products are more likely when the product has marketplace advantages, is targeted at an attractive market, and is well executed through excellent internal organization" (p. 352). In general, rational plan research offers "an early and a broad understanding of which factors are essential for successful product development and for emphasizing the role of the market in what is often conceived of as a purely technical or organizational task" (p. 353).

"The underlying premise [of product development as communication web] is that communication among project team members and with outsiders stimulates the performance of development teams. Thus, the better that members are connected with each other and with key outsiders, the more successful the development process will be" (pp. 353-354). As widely focused as the rational path focus is, communication web "is narrowly focused on one independent variable—communication, [emphasizing] depth, not breadth as in the rational plan, by looking inside the "black box" of the development team. The authors find the communication web perspective to be complementary to rational plan, in that the latter includes "political and information-processing aspects of product development. The result is excellent theoretical understanding of a narrow segment of the phenomenon. In this case, there also is greater methodological sophistication (e.g., multiple informants, multivariate analysis) than in the first stream" (p. 354). Associated with the communication web perspective is a "typology of external communication or 'boundary-spanning' behaviors. Ambassador activities consisted of political activities such as lobbying for support and resources as well as buffering the team from outside pressure and engaging in impression management. Task coordination involved coordination of technical or design issues. Scouting consisted of general scanning for useful information, whereas guard activities were those intended to avoid the external release of proprietary information" (p. 356).

Another surprising finding "was that the frequency of external communications was not a significant predictor of team performance. Rather, communication strategy was germane. The most successful product-development teams engaged in a comprehensive external communication strategy, combining ambassador and task-coordination behaviors that helped these teams to secure resources, gain task-related information, and so enhance success" (p. 356). The fact that "more effective teams engaged in both political and task-oriented external communications [suggests] that product-development teams must attend not only to the frequency of external communication, but also to the nature of that interaction" (p. 357). Gatekeepers, those "who encourage team communication outside of their groups, and powerful project managers, who communicate externally to ensure resources for the group," facilitate product development by having their "teams also engage in extensive political and task-oriented external communication. The underlying rationale is that politically oriented external communication increases the resources of the team, whereas task-oriented external communication increases the amount and variety of information. These types of communication, in turn, aid the development-process performance" (p. 358).

The third research perspective is "disciplined problem-solving" which sees "successful product development...as a balancing act between relatively autonomous problem solving by the project team and the discipline of a heavyweight leader, strong top management, and an overarching product vision." The literature of this perspective has established that "strong formal ties to suppliers and R&D networks were very important to the product-development process. In such networks, suppliers can acquire a very high level of technical skill in a specialized area, which allows them to fulfill sudden or unusual requests quickly and effectively" (p. 359). The more varied the functional specializations of team members, the greater access to diverse information the product development team has. Cross-functionality of team members facilitates the easy overlapping of development phases, "which also quickened the pace of product development. Furthermore...product development was accelerated by overlapping of development phases and cross-functional teams only if supported by continuous communication among project members" (p. 362).

The final concept that Brown and Eisenhardt put forth as a result of their findings is senior management's engagement in "subtle control." This can occur when "members of successful project teams maintain a balance between allowing ambiguity, such that creative problem solving...at the project team level [allows], and exercising sufficient control, such that the resulting product fits with overall corporate competencies and strategy." Senior management can engage "in subtle control by communicating a clear vision of objectives to their teams while simultaneously giving team members the freedom to work autonomously within the discipline of that vision" (p. 362).

CONCLUSION

Much of the consideration given to new product development follows two different themes: understanding success vs. failure and the process of product development. In addition to these themes, consideration of other critical issues was presented by Teece (1986), Mansfield (1986), Mahajan et al. (1990), and Brown and Eisenhardt (1995).

Clearly understanding what separates a successful new product from a failure is critical to the reduction and elimination of failed products through application of best practices and understanding the theoretical foundations of successful product development. Project scope has been found to have a significant effect on product development, one that helps explain differences between the performance of Japanese and U.S. firms. According to Clark (1989), scope has two elements: (1) the use of unique vs. off-the-shelf parts, and (2) the nature of supplier involvement.

In new product performance studies, there is a wide variation in results that are surprisingly non-convergent (Montoya-Weiss & Calantone, 1994). In general, success studies identify characteristics and factors leading to success, whereas failure studies provide retrospective analyses of past failures to identify the determinants of failure and common pitfalls or problems in the development process. In other words, success studies provide less detail and explanation as to why and how new products are successful. Having warned of the contextual issues associated with success and failure studies, the results of four seminal studies are offered.

Cooper and Kleinschmidt (1986) found the following factors to be fundamental to new product success:

1. Differential advantage;
2. A market orientation including an understanding of users' needs, wants, and preferences;
3. A strong product launch;
4. Technological strengths and synergy with resources and skills of the firm;
5. Marketing synergy with the product;
6. An attractive market; and
7. Top management support and commitment.

In general, new product success is closely linked to the activities that are carried out in the new product process, their execution, and the completeness of the process.

A comparison of product and development success and failure in theory and practice (Griffin & Page, 1993) found:

- S/F is multidimensional.
- Five independent dimensions of S/F performance are: firm-, program-, and product-level measures, and measures of financial performance and customer acceptance.
- Practitioners use about four measures from a total of two different dimensions, most frequently customer acceptance and financial permanence.
- Researchers use fewer measures, about three, from one to two dimensions. The particular dimensions used differ across three different clusters of researcher focus.
- Researchers focus more on overall firm impacts of S/F.
- Companies focus on the S/F of individual projects.

Cooper (1979) finds that factors that differentiate new product success and failure for success include: (1) product uniqueness and superiority, (2) market knowledge and marketing proficiency, and (3) technical and production synergy and proficiency. Factors for failure include: (1) high-price relative to competition, (2) being in a dynamic market, and (3) being in a competitive market where customers are already well satisfied. When differentiating between new product success and failure, product superiority is the number one factor influencing commercial success; project definition and early predevelopment activities are the most critical steps in the new product development process. Characteristics identified as the most important discriminators between commercial successes and failures are (Cooper & Kleinschmidt, 1987, p. 171):

1. Understanding of users' needs,
2. Attention to marketing and publicity,
3. Efficiency of development,
4. Effective use of outside technology and external scientific communication, and
5. Seniority and authority of responsible managers.

Having summarized the seminal work that considers understanding success vs. failure in new product development, the process of product development is given further consideration. Planning, cooperation, and speed are the foci of improving the product development process.

The value of developing an aggregate project plan to plan and monitor development is offered (Wheelwright & Clark, 1992). An aggregate project plan categorizes projects based on the amount of resources they consume and on their contribution to the company's product line. By mapping the project types, gaps in the development strategy can be identified, and informed decisions regarding what types of projects to add and when to add them can be made. Activities should be classified based on the degree of change in the product and the degree of change in the manufacturing process.

Guidelines to support and assist in improving the interface between R&D and marketing, a critically important relationship for development of new products, are:

1. Break large projects into smaller ones,
2. Take a proactive stance toward interface problems,
3. Eliminate mild problems before they grow,
4. Involve R&D and marketing early in the life of the project,
5. Promote and maintain dyadic relationships,
6. Open communication is everyone's "responsibility,"
7. Use interlocking task forces, and
8. Clarify the decision authorities and responsibilities.

These points all stress the importance of the marketing and R&D interface to the development of new products (Souder, 1988). The continuing pressure to develop products more quickly led Millison et al. (1992) to offer a series of tactics to accelerate new product development. These tactics are: (1) simplify, (2) eliminate delays, (3) eliminate steps, (4) speed up operations, and (5) parallel process.

Other important insights are given regarding the benefits and pitfalls of being a first mover, the use of intellectual property and patents for new products, new product diffusion models, and the consideration of product development through different perspectives. Consideration is given to the failure of first mover advantage to always bear benefit to the developers of new products. Teece (1986) suggests that when imitation is easy, markets do not work well, and the profits from innovation may accrue to the owners of certain complementary assets, rather than to the developers of the intellectual property. The use of patents resulted in a discovery that industry is a critical factor to the determination of whether inventions should be patented or not (Mansfield, 1986). Refinements to new product diffusion models to address different important questions and context issues are identified as critical (Mahajan et al., 1990), speaking to both the importance and limitation of Bass's (1969) work on new product diffusion.

Additional insights into new product development can be offered through the lens of different perspectives. Brown and Eisenhardt (1995) identify critical economic and organizational perspectives. They further divide product development into: rational plan, communication's web, and disciplined problem solving. Having considered new product development, the management of communication and information technology is now addressed.

REFERENCES

Brown, S.L., & Eisenhardt, K.E. (1995). Product development—past research, present findings and future directions. *Academy of Management Review, 20*(2), 343-378.

Clark, K.B. (1989). Project scope and project performance—the effect of parts strategy and supplier involvement on product development. *Management Science, 35*(10), 1247-1263.

Cooper, R.G. (1979). Dimensions of industrial new product success and failure. *Journal of Marketing, 43*(3), 93-103.

Cooper, R.G., & Kleinschmidt, E.J. (1986). An investigation into the new product process—steps, differences and impact. *Journal of Product Innovation Manage-*

ment, 3(2), 71-85.

Griffin, A., & Page, A.L. (1993). An interim report on measuring product development success and failure. *Journal of Product Innovation Management, 10*(4), 291-308.

Mahajan, V., & Muller, E. (1979). Innovation diffusion and new products growth models in marketing. *Journal of Marketing, 43*(4), 55-68.

Mahajan, V., Muller, E., & Bass, F.M. (1990). New product diffusion models in marketing—a review and directions. *Journal of Marketing, 54*(1), 1-26.

Mansfield, E. (1986). Patents and innovation—an empirical study. *Management Science, 32*(2), 173-182.

Millison, M.R., Raj, S.P., & Wilemon, D. (1992). A survey of major approaches for accelerating new product development. *Journal of Product Innovation Management, 9*(1), 53-69.

Montoya-Weiss, M.M., & Calantone, R. (1994). Determinants of new product performance: A review and meta-analysis. *Journal of Product Innovation Management, 11*(5), 397-417.

Rogers, E. (1983). *Diffusion of innovations.* New York: The Free Press.

Souder, W.E. (1988). Managing relations between R&D and marketing in new product development projects. *Journal of Product Innovation Management, 5*(1), 6-19.

Teece, D.J. (1986). Profiting from technological innovation: Implications for integration, collaboration, licensing and public policy. *Research Policy, 15*(1), 285-305.

Wheelwright, S.C., & Clark, K.B. (1992). Creating project plans to focus project development. *Harvard Business Review, 70*(2), 70-82.

Section III
Technology and Management Information Systems: Executive Summary

In this section, we emphasize the powerful developments in research interests and approaches to computing technology itself, for in aggregate, computers and networks are increasingly integral to a firm's innovative ends and operative strategies. We begin with a look at how individual predispositions to computerized information systems affect human performance with them, and investigate information processing and decision behavior, and their effect on the successful development of an organization's management information systems. Researchers also look at MIS users' expectations prior to the systems' implementation to see how and if these expectations could aid the MIS development process. The nature of change regarding end user involvement in information and communication technologies from the mid-1980s through the current time is discussed, as is user involvement and its relationship to system success. The well-known Theory of Reasoned Action (TRA) and the Technology Acceptance Model (TAM) are applied to the question as to why people accept or reject IT. Suggestions as to why there is a positive correlation between product design and customer need include that inter-departmental collaboration is essential to IT product development, and that innovators do not necessarily seek out the benefits of such collaboration when tying technological and marketplace issues together as their products are developed are offered, as are survey results from 185 top IS executives regarding their views on the types of applications that augment their competitive advantage. Information systems development is approached from a strategic and organizational (as opposed to a user-based) vantage point through adaptive structuration theory (AST), a framework to be implemented when examining organizations for signs of change derived from using IT. The chapter concludes with a discussion of the Technology Acceptance Model, but with consideration given to additional variables and their effects on perceived usefulness and perceived ease of use of information technologies.

The articles addressed in Chapter X provide insight into the challenges that software development poses to those who follow a proprietary path as well as specific situations in which management and innovation theory is responsive also to non-proprietary, or open source software development. Subjects discussed include a theory to aid software project management, and risk management as it pertains to software development projects; knowledge management in software process innovation management environments; the mutually advantageous bodies of knowledge that the realms of software development and new product development hold for one another; software development management from a knowledge and team management perspective; and e-knowledge management systems and their application to new product development. The articles addressed in the second part of this chapter have particular relevance to software's place in the wider scope of technology innovation management. Issues include the open source movement's reshaping of the private investment and the collective action models of innovation; the rationale of programmers who invest themselves, without remuneration, in software development; why some programmers assist proprietary firms engaged in software development projects; and ways to merge proprietary and open source methodologies and management styles with established theories of appropriability and adoption as they pertain to software development.

Chapter IX
Information and Communication Technology Management

INTRODUCTION

In this chapter on information and communication technology management, we retain a chronological order to emphasize the development of research interests and approaches as technology itself grows more complex, sophisticated, and increasingly integral to a firm's innovative ends and operative strategies. We begin with two articles concerned with behavior—specifically, attitudes and decision behavior in the early realm of management information systems. Robey (1979) looks at the attitudes of members of sales departments to understand how individual predispositions to computerized information systems affect human performance with them, ultimately suggesting that the identification of expectancy factors can coalesce into a model of user reactions and motivations toward MIS. In the same year, Zmud investigates information processing and decision behavior, and their effect on the successful development of an organization's management information systems.

In the 1980s, the focus shifts away from behaviors and toward methodologies and practices of MIS development and their implementation. Similarly to Zmud (1979), Ginzberg (1981) presents the results of his study involving MIS users' expectations prior to the systems' implementation to see how and if these expectations could aid the MIS development process. McFarlan's 1984 article addresses CEOs, primarily, urging them to be aware of the double-edged nature of MIS, as these systems can both create and preclude entry into new domains. That same year, Parsons offers a tri-level framework that helps senior management determine the effectiveness of information technologies and the prospective impacts that they might have on their organizations.

It is fitting that theory and implementation of strategies regarding end user involvement is the third and concluding section of this chapter, as the progression of information and communication technologies from the mid-1980s through the current time has shifted focus to the individual—as customer, as development partner, and as arbiter of product design and modeling. Ives and Olson's 1984 article provides a review of the research on user involvement and its relationship to system success. Stefik et al. (1987) provide an early view of computer-assisted collaborative work tools. Davis, Bagozzi, and Warshaw (1989) look toward the Theory of Reasoned Action (TRA) and the Technology Acceptance Model (TAM) to discuss when and why people accept or reject IT. Melone (1990) provides a comprehensive assessment of a user satisfaction construct, looking at how end user attitudes function as a factor of it. Dougherty (1992) offers her interpretation of both innovation literature and innovators' practice with the marketplace from a philosophical vantage point. She suggests that there is a positive correlation between product design and customer need, that inter-departmental collaboration is essential to IT product development, and that innovators do not necessarily seek out the benefits of such collaboration when tying technological and marketplace issues together as their products are developed. Sethi and King (1994) supply survey results from 185 top IS executives regarding their views on the types of applications that augment their competitive advantage. DeSanctis and Poole (1994) approach information systems development from a strategic and organizational (as opposed to a user-based) vantage point with their adaptive structuration theory (AST), a framework to be implemented when examining organizations for signs of change derived from using IT. Orlikowski (1996) introduces situated practice as a methodology to understand the relationships between organizational change and IT, and an alternative to established perspectives such as planned change, technological imperative, and punctuated equilibrium. Venkatesh and Davis (2000) conclude this chapter with a discussion of Davis's Technology Acceptance Model, but with consideration given to additional variables and their effects on perceived usefulness and perceived ease of use of information technologies.

Attitudinal Responses to IT and MIS Implementation

Robey (1979) studies attitudes displayed by an industrial sales force to understand how certain predispositions affect the use of computer-based information systems.

He begins with the premise that researchers should not spend any time "reinvent[ing] theory and learn[ing] empirical lessons through their own mistakes rather than through the experience of others," and suggests a new "expectancy-based model of user reaction [and job motivation] to MIS" (p. 528).

The fundamental problem with MIS, in Robey's assessment, is system designers' not factoring in users' psychological and behavior responses to a system, and/or the organizational culture in which the MIS is to be used. Robey's findings support "the established notion that user attitudes (or perceptions) are significant correlates of system use. Attitudes are less powerful in predicting subjective assessments of perceived worth, although the relationships are significant....The finding that attitudes are more strongly related to actual use than they are to measures of perceived worth has important implications. If it is assumed that MIS designers and managers are more interested in actual usage of MIS, it seems important to focus on actual use in research" (p. 534).

Information Processing

Zmud's 1979 article reports on his multidisciplinary synthesis of research findings concerning information processing and decision behavior, and their effect on the successful development of an organization's management information system (MIS). Researchers usually address issues such as "organizational characteristics, environmental characteristics, task characteristics, personal characteristics, interpersonal characteristics, MIS staff characteristics, and MIS policies," influencing "individual differences upon MIS design, implementation, and usage" by "examining the effect of a number of cognitive, personality, demographic, and situational variables upon information processing and decision behavior" (Zmud, 1979, p. 966).

Zmud finds that the "individual differences believed most relevant to MIS success are grouped into three classes" of variables: "cognitive style, personality, and demographic/situational" (p. 967). He finds that the empirical literature on cognition, or "the activities involved in attempts by individuals to resolve inconsistencies between an internalized conceptualization of the environment and what is perceived to be actually transpiring in the environment" (p. 968), clearly indicates that individual differences in this variable "do exert a major force in determining MIS success" (p. 975).

Methodologies and Practices of IT and MIS Implementation

Ginzberg (1981) presents the results of his study into "the initial exploration of MIS user pre-implementation expectations, to determine if they could serve as the basis for a tool to help guide and manage the MIS development process" (p. 474).

No doubt, management and all systems users would benefit by understanding how they can contribute to the systems development process, but unlike Robey, whose focus is on user behaviors, attitudes, and measuring user and organizational characteristics, Ginzburg's focus is on systems developers and tools they can use to increase the effectiveness of the systems they build. What are the "variables [that] are both particularly important to successful system development and controllable by the system designer or the user," and how can the development process benefit from the development of tools that "designers and users can employ to manage" that process? (p. 460).

Ginzberg posits that the behavioral approach may not be as effective and productive as providing "guidance for the management of ongoing implementation efforts" in terms of systems development. His study is a step toward discovering a way "to address the implementation management question by exploring the use of MIS users' pre-implementation expectations about a system as indicators of the likely success of that system." He approaches system development as a series of stages that begin with a definition of the functions that are intended to comprise the system, and in which "most of the key decisions about the system as the user will see it are made." With only a quarter of the total resources required for system development being expended at this initial stage, "the decisions, which will have the greatest effect on the users' acceptance or rejection of a system, are made prior to the bulk of spending on the project, and an assessment of the project's probability of success or failure should be possible at that time" (p. 459).

Failure to implement occurs most often when "users hold unrealistic expectations about a system." Therefore, Ginzberg suggests identifying user expectations by the end of the definition stage, as they "might serve as early warning indicators of MIS implementation outcomes." If reliable, the indicators would serve as diagnostics, and suggest changes in direction and outcome early in the project. The results of his study indicate that users who have a realistic sense of what the system can do before it is even implemented will be generally more satisfied with that system than those who hold unrealistic goals for it. Ginzberg calls for more research into ways to identify and define expectations, "simpler tools for measuring expectations, [and] proper timing of expectations measurement before reliable instruments for measuring expectations in ongoing projects" (p. 459).

Ginzberg suggests working from G.B. Davis's formulation of the stages of system development found in his 1974 *Management Information Systems: Conceptual Foundations, Structure and Development,* where Davis calls for understanding the development process as being composed of three main stages, with refined sub-stages occurring in each. However, no matter how many stages a particular model presents, one must recognize that "the activities taking place are essentially different from stage to stage. Furthermore, for most stages, the activity of the preceding stage must be completed before the new stage can be started" (Ginzberg, 1981, p. 460).

The first stage, "Definition," establishes the functionality and (now in the age of the graphical user interface) how the new system presents itself to the user. Activities in the first stage include feasibility assessment and information analysis. Stage Two—Physical Design—builds on the specifications produced in Stage One, with the goal of meeting those specifications. Stage Two consists of the sub-stages of system design, program development, and procedure development. Lastly, the Implementation Stage, which consists of conversion, operation and maintenance, and post-audit functions, is when "the physical system is installed, operated, and monitored" (p. 461).

A great deal is riding on the proper management of user expectations. Research directions that are beneficial to the goal of reducing instances of failure include areas such as the effects of "disconfirming expectations on performance satisfaction...product evaluation...and job satisfaction" (p. 462). Ultimately, when users have a realistic sense of what they can expect from a management information system before the design is implemented, the project outcome, in terms of users' attitudes and behaviors, will be positive. Ginzberg suggests that system developers play an active role during the first stage, as a way to provide "benchmark expectations against which the realism of users' expectations can be judged" (p. 475).

Assessing Competitive Impact of IT

Parsons' 1983 *Sloan Management Review* article "presents a three-level framework to help senior managers assess the current and potential impact of IT on their businesses." Developed "from the results of a two-year study of more than a dozen companies...[it] is based on a recognition and analysis of the competitive environment and strategies of business, and focuses on the opportunities for firms to use IT to improve their competitive positions" (p. 4). Recognizing the extent and rapidity of change that information technologies had introduced to many industries by the early 1980s, Parsons' aim is to explain why, even though "many managers understand the potential impact of information systems on their firm's competitive position, others fail to consider strategy implications when selecting a system." Parsons' contribution is his "multilevel framework for assessing the competitive impact of information technology on a firm, and [provision of] a guide for integrating information systems with a firm's strategy" (p. 3).

Parsons finds that there are three levels of impact that information technology has on businesses. Industry-level impacts include "Products & Services; Markets; [and] Production Economics." At the firm level, IT affects key competitive forces such as "Buyers; Suppliers; Substitution; New Entrants; [and] Rivalry." There is also a strategy level, at which IT affects "Low-Cost Leadership; Product Differentiation; [and] Concentration on Market or Product" (p. 4). Even though there is a "general lack of strategic direction, firms in many industries are using IT to their competitive advantage. As technology continues to evolve rapidly and becomes a major factor in more industries, firms must begin to strategically manage information technology." Parsons' framework "describes the first steps management should take to link IT to a firm's competitive environment and strategy by analyzing the impact of IT at the industry, firm, and strategy levels" (p. 13).

User Involvement in IT and MIS Implementation

Ives and Olson's 1984 article is a review of research on user involvement and its relationship to system success. Generally understood to be a necessary component of evaluating information systems, user involvement as a subject of study suffers from a paucity of quantitative evidence and plausible theory supporting this assumption. The authors conducted a comprehensive review of literature that "examines the link between user involvement and indicators of system success" (p. 586), only to find that previous researchers' experimental methodologies were flawed and/or their findings ill fitting to established and accepted theory. To rectify this situation, Ives and Olson offer, in addition to the literature review, a theoretical framework in which to consider the relationship between user involvement and computer information system success.

When Ives and Olson refer to user involvement, they mean end users' participation in the development of computer information systems. Systems developers who seek this participation do so with the operative idea that the input of target users increases the chances of system success. This belief can be tracked back to "theory and research in Organizational Behavior, including group problem-solving, interpersonal communication, and individual motivation" (p. 587). The authors make the point that in the literature there is no overt statement linking organization behavior theory to research in user involvement; nevertheless, they identify strong positive correlations between the methods and findings of the two bodies of literature and sets of theories.

For instance, Ives and Olson look toward the organizational behavior notion of "participative decision-making (PDM)" as a credible parallel theory to follow when assessing the value of end user participation in system development. From that point of view, PDM "is to increase inputs of subordinates into management decisions that are related to their jobs" (p. 587). Previous research has predicted that user participation will have a positive impact on system quality because end users can provide an accurate and comprehensive assessment of requirements; offer expertise about the organizational context in which the system is to be implemented; point out, from the end user perspective, extraneous features of the developing system; and assist other prospective end users understand the system.

The authors also discuss the overlay of the organizational behavior theory of "planned organizational change implementation," which considers the success of a computer information system to be understood as that system's acceptance and use by its intended users. Those promoting planned organizational change implementation perceive system success to be dependent on the quality of the implementation process (Ives & Olson, 1984, p. 588). The system itself is viewed as a change agent, one promoting individual attitudinal change, which results in organizational change. However, proponents of this view also see end user involvement as being a necessary, but not sufficient "condition for decreasing resistance and increasing acceptance of the change." Investigators of planned change view it as either the result of collaboration or negotiation between managers and end users, suggesting that positive outcomes accrue from "open interaction and joint evaluation" (p. 588).

In the end, however, after Ives and Olson review the relevant organizational behavior literature, they come to the conclusion that there is only a weak connection between established theory and the premises that researchers use in designing their research questions on user involvement. Empirical research studies are also insufficient in design and results to demonstrate any benefit to user involvement in the computer information system design process. Ultimately, "not only has empirical research been unable to foresee when and what types of user involvement are appropriate, it has not convincingly demonstrated that user involvement contributes to system success" (p. 601).

Computer Supported Collaborative Work

Computing technologies have long been employed to assist our managing of information and decision making. Stefik et al. (1987) present an early study of computer-supported collaborative work at Xerox PARC through descriptions of the activities and technologies in their Colab, where "computers support collaborative processes in face-to-face meetings." Even though a substantial portion of our workday involves meetings, the passive tools used to help participants reach decisions and conclusions are not the tools we as individuals are practiced in using when working individually. Like all media, computer systems "influence the course of a meeting because they interact strongly with participants' resources for communication and memory." Unlike flip-charts and chalkboards, which are unreliable storage devices, computers offer a variety of aids that "provide flexibility for rearranging text and figures...file systems...to retrieve information generated from previous meetings, revisit old arguments and to resume discussions" (p. 32).

The Colab experimental meeting room consisted of networked, touch-screen computers running software that draws data from a distributed database, allowing real-time editing/updating access for all participants to documents and graphics.

The authors "describe the meeting tools we have built so far as well as the computational underpinnings and language support we have developed for creating distributed software." A basic requirement for any meeting tool is provision of "a coordinated interface for all participants" to facilitate interaction among users "through a computer medium." Each participant, having access to the same tools, is operating in a WYSIWIS environment—"what you see is what I see," referring to "the presentation of consistent images of shared information to all participants. A meeting tool is strictly WYSIWIS if all meeting participants see exactly the same thing and where the others are pointing. WYSIWIS creates the impression that members of a group are interacting with shared and tangible objects [and] recognizes the importance of being able to see what work the other members have done and what work is in progress: to 'see where their hands are'" (p. 33).

The authors created three tools that "support simultaneous action," facilitating efficiency. A "key issue in tool design is recognizing and supporting those activities that can be decomposed for parallel action. For parallel action, a task must be broken up into appropriately sized operations that can be executed more or less independently by different members of the group." However, the "ability to act in parallel on shared objects also brings with it potential for conflict." The designers, therefore, created a "conflict detection system or 'busy signal' [that] graphically warns users that someone else is already editing or otherwise using an item." Cognoter, one of the tools created, "closely imitates the functionality of a chalkboard. It is intended for informal meetings that rely heavily on informal freestyle sketching....Other Colab tools are based on much more formal models of the meeting process." The authors present for consideration "Cognoter, a tool for organizing ideas to plan a presentation; and Argnoter, a tool for considering and evaluating alternate proposals" (p. 34).

Cognoter's "output is an annotated outline of ideas and associated text." It is "unique in that it is intended for collective use by a group of people [organizing] a meeting into three distinct phases—brainstorming, organizing and evaluation" (p. 34). In the brainstorming phase, "ideas are not evaluated or eliminated...and limited attention is given to their organization....A participant selects a free space in a public window and types in a catchword or catchphrase characterizing an idea." When in the organizing phase, "the group attempts to establish an order for the ideas generated in the brainstorming phase. With Cognoter, the order of ideas can be established incrementally by using two basic operations: linking ideas into presentation order and grouping ideas into subgroups." Finally, in the evaluation phase, participants "review overall structure and reorganize ideas, fill in missing details, and eliminate peripheral and irrelevant ideas. In Cognoter, the various decision-making processes are separate and distinct operations." Cognoter "helps in the final ordering of ideas by preparing an outline and indicating which ideas are ordered arbitrarily...divides the thinking process into smaller and different kinds of steps that are incremental and efficient [and provides the] transitivity and grouping operations [that] make it possible to organize the ideas efficiently with a small number of links" (pp. 35-37).

Argnoter is "an argumentation spreadsheet for proposals....Proposal meetings start when one or more members of the group have a proposal for something to be done, typically a design for a program or a plan for a course of research. The goal of the meeting then becomes to pick the best proposal." Given that "discovering, understanding and evaluating disagreement are...essential parts of informed decision making in these meetings...the intuition guiding the Argnoter process is the recognition that much of the dispute and misunderstanding that arise in meetings about design proposals are due to three major causes: owned positions, that is, personal attachment to certain positions; unstated assumptions; and unstated criteria." These causes promulgate technical responses through Argnoter, with its use in meetings comprising, like Cognoter, three distinct phases: "proposing, arguing, and evaluating." In the proposal phase, "a proposal will be created in, and displayed by, a set of connected windows called proposal "forms," which can be either private or public. Proposal forms are WYSIWIS, whereas a private form appears only on the machine of the participant who controls it." In the arguing phase, participants "write pro and con statements about all proposals, not just pro statements for the ones they are in favor of and con statements for the rest" (p. 38). During the evaluation stage, Argnoter helps meeting participants consider "the assumptions behind individual arguments," with assumptions considered to "statements about statements," by modeling "differences with explicit 'belief sets,' [or] a mapping of a set of statements into valid (believed) or invalid (not believed) categories....The act of making belief sets explicit enables Argnoter to act as a kind of argumentation spreadsheet where a proposal is viewed and evaluated in relation to a specified set of beliefs" (p. 39).

The authors provide observations and future research questions deriving from the creation of their meeting tools, beginning with the relationship between Cognoter and the writing process. While the "current Cognoter design reflects a set of conjectures regarding the writing process, from the early stage of idea generation and development through the generation of a path or outline for a final presentation...the actual use of Cognoter revealed not only the points of fit between design and process, but some subtle disjunctures as well....In early sessions with Cognoter, we found that even before moving on to the organizing phase, members began using spatial grouping in the brainstorming window to display relationships between ideas. Even after items were explicitly linked, the spatial cues helped to display the relationships between items; these spatial cues, in turn, were important to the elaboration of meaning" (pp. 43-44). Given that "the Colab's starting premise was that serial access to problem-solving technology obstructs the kind of equal participation that ideally characterizes collaboration," the designers attempted to equalize "access of all participants to displays and shared data [through] an interface [that] enhances flexibility as to roles and discourages control over the activity by any one participant." Even so, "our early sessions demonstrated that the constraints imposed by current technologies are not just a limitation on collaboration but in some ways a resource as well" (p. 44).

The author's "guiding question has been, What are the processes of collaboration for which the computer is an appropriate tool, and what particular Colab tools could be designed to support these processes?...Cognoter takes a joint presentation as its object and encourages consensus by supporting a single viewpoint, whereas Argnoter encourages competing proposals and delayed consensus by allowing the display and comparison of multiple views." Future research includes studies seeking answers to questions such as: "To what extent do our work practices compare and contrast with other settings and other participants? Does a tool, by reifying a process and making it explicit, thereby also make it portable across groups? Or do we need a set of tools that can be customized to different users in different settings? Under what circumstances are explicit structures desirable, and under what circumstances do we want to minimize the amount of structure we build into our tools?" (p. 45). These are questions to which, since 1987, the professions and disciplines of human-computer interaction, artificial intelligence systems, and mediated communications technology development have been applied.

Management Views on IT and MIS Implementation

McFarlan (1984), writing in the *Harvard Business Review* decades before the Internet moved the global economy into the information age, posits that information systems and information technology in general are considered supporting characters in business management. "This article offers a discussion about the strategic resource, which technology, computers and information systems" represented over 20 years ago, and suggests that CEOs pay attention to the "competitive advantage" these systems provide, particularly because they can build or "counter barriers to entry." Moreover, companies can use information systems and technology can be helpful in building in "switching costs, and even, sometimes, to completely [changing] the basis of competition." Rather than playing "a difficult and expensive game of catch-up ball," some savvy executives conduct competitive analysis in assessing where IS fits into their companies, since "in some cases it appropriately plays a support role and can add only modestly to the value of a company's products, while in other settings it is at the core of their competitive survival [in order to] determine both the proper level of expenditures and the proper management structure for IS" (p. 98).

Given that, as is often the case, "the new technology has opened up a singular, one-time opportunity for a company to redeploy its assets and rethink its strategy," information technologies have provided "potential for forging sharp new tools that can produce lasting gains in market share" (p. 99). However, new systems of information retrieval and analysis now also require "management to change the way it operates." Specifically, McFarlan states:

1. "The CEO must insist that the end products of IS planning clearly communicate the true competitive impact of the expenditures involved." In 1984, computing equipment and software expense in relation to information return were far greater than they are today, demanding strategic investment for achievable goals.
2. "Till now, it has been the industry norm for organizations and individuals to share data widely about information systems technology and plans, on the ground that no lasting competitive advantage would emerge from IS and that collaboration would allow all to reduce administrative headaches. But today, managers should take appropriate steps to ensure the confidentiality of strategic IS plans and thinking." The strategic advantages to unique uses of information systems were viewed as far too valuable to support distribution of general knowledge about them.
3. "Executives should not permit the use of simplistic rules to calculate desirable IS expense levels." That is, the potential return on investment in information systems is very difficult to calculate, given the potential additional uses and benefits of the systems acquired.
4. "Inter-organizational IS systems have hidden, second-order effects, that is, repercussions in other parts of the business." While some of these effects may be productive, there are situations that may prove to be negative for some divisions or departments.
5. "Managers must not be too efficiency-oriented in IS resource allocation." Given the potential benefits attendant to data mining and other information strategies not presented as the explicit use of a particular system, non-IS groups may prove to be effective users of IT (pp. 102-103).

Socio-Psychological Theory and IT/MIS Implementation

Davis et al. (1989) examine "the ability of theory of reasoned action (TRA) and technology acceptance model (TAM) to predict and explain user acceptance and rejection of computer-based technology." Specifically, they are "interested in how well we can predict and explain future user behavior from simple measures taken after a very brief period of interaction with a system. [A] longitudinal study of

107 MBA students...provides empirical data for assessing how well the models predict and explain voluntary usage of a word processing system." Coupled with a discussion of "the major characteristics of the two models [and] the prospects for synthesizing elements of the two models," Davis et al.'s goal is to "arrive at a more complete view of the determinants of user acceptance" (p. 983).

Davis et al. cite Ajzen and Fishbein (1980) and Fishbein and Ajzen (1975) to offer a definition of TRA as "a widely studied model from social psychology...concerned with the determinants of consciously intended behaviors...According to TRA, a person's performance of a specified behavior is determined by his or her behavioral intention (BI) to perform the behavior, and BI is jointly determined by the person's attitude (A) and subjective norm (SN) concerning the behavior in question...with relative weights typically estimated by regression: BI is a measure of the strength of one's intention to perform a specified behavior...A is defined as an individual's positive or negative feelings (evaluative affect) about performing the target behavior...subjective norm refers to the person's perception that most people who are important to him think he should or should not perform the behavior in question" (Davis et al., 1989, pp. 983-984).

"TAM, introduced by Davis (1986), is an adaptation of TRA specifically tailored for modeling user acceptance of information systems. The goal of TAM is to provide an explanation of the determinants of computer acceptance that is general, capable of explaining user behavior across a broad range of end-user computing technologies and user populations, while at the same time being both parsimonious and theoretically justified" (p. 984). Although contemporary readers may accept computer use as a given in daily life, Davis et al. point out that, a generation ago, "resistance to end-user systems by managers and professionals is a widespread problem. To better predict, explain, and increase user acceptance, we need to better understand why people accept or reject computers," hence their study (p. 982).

Davis et al. "address the ability to predict peoples' computer acceptance from a measure of their intentions, and the ability to explain their intentions in terms of their attitudes, subjective norms, perceived usefulness, perceived ease of use, and related variables, [through] a longitudinal study of 107 users, [in which] intentions to use a specific system, measured after a one-hour introduction to the system, were correlated 0.35 with system use 14 weeks later." At the end of the test period, the "intention-usage correlation was 0.63," suggesting that "perceived usefulness strongly influenced peoples' intentions...Perceived ease of use had a small but significant effect on intentions as well, although this effect subsided over time. Attitudes only partially mediated the effects of these beliefs on intentions" (p. 982).

In keeping with the pervasive attitudes regarding computing and management information systems of the time, Davis et al. acknowledge that "organizational investments in computer-based tools to support planning, decision-making, and communication processes are inherently risky. Unlike clerical paperwork-processing systems, these 'end-user computing' tools often require managers and professionals to interact directly with hardware and software" (p. 982). It seems obvious, in hindsight, that "practitioners and researchers require a better understanding of why people resist using computers in order to devise practical methods for evaluating systems, predicting how users will respond to them, and improving user acceptance by altering the nature of systems and the processes by which they are implemented" (p. 984). Davis et al. find that "subjective norms had no effect on intentions [and the] results suggest the possibility of simple but powerful models of the determinants of user acceptance, with practical value for evaluating systems and guiding managerial interventions aimed at reducing the problem of underutilized computer technology" (p. 982).

User Satisfaction and IT Effectiveness

In a highly cited and relatively recent *Management Science* article, Melone (1990) "assesses the user-satisfaction construct from a theoretical perspective, focusing on the function of a user's attitude, its structure, and its relationship with other constructs" (p. 77). Unlike previous research in this area that has roots in social and cognitive psychology, Melone's paper looks into why user satisfaction is a popular evaluation construct and what assumptions academicians and practitioners have for believing it to be useful in measuring effectiveness.

User satisfaction is a bedrock construct for behavioral research in the academic discipline of Information Systems and is a metric industry uses to gauge IS effectiveness. Melone's goal is to provide a theoretical assessment of user satisfaction as a foundational idea, "by examining conceptual and theoretical limitations of the construct's use [with a focus] on the evolution of the construct in the literature and the theoretical problems associated with its broader use" (p. 76). He finds similarities between user satisfaction and behavioralist definitions of "attitude." Research exists that explores "alternative theoretical views on attitude structure that parallel research findings on user satisfaction, specifically expectancy-value models...the family of cognitive approaches" is distinct from theories on user satisfaction. The author considers how these two attitude structures can benefit research in Information Systems, and discusses "the ways in which these structures have been integrated in terms of understanding the relationship of users' active responses to other responses (i.e., behavior or cognition)" (p. 76).

Melone's findings include the point that when "measuring effectiveness, user satisfaction alone is not sufficient to adequately capture the full meaning of effectiveness"; that "in addition to affect-oriented measures of effectiveness, output-oriented criteria should be employed, [such as] the cognitive (arguments and information) and behavioral (actions and intentions) components of attitude" and that "attitudes are considerably more complex than conceptualized in most user-satisfaction research" (p. 88).

Interpretive Schema and IT Innovation

Dougherty (1992) presents a reasoned and reasonable interpretation of both innovation literature and innovators' practice with the marketplace, and does so from a quasi philosophical perspective. The paper presents three main findings. First, product success is positively correlated to the closeness in fit between product design and customer need. Second, the more collaboration among departments within a large corporation producing an innovative product, the greater the chances are of the product's success. Third, those who innovate often do so without the benefit of inter-departmental collaboration or overtly linking the technological and market issues involved in their product.

Beginning with the premise that the "development of commercially viable new products requires that technological and market possibilities are linked effectively in the product's design," Dougherty readily accepts that large corporations find such linkages problematic. Her article focuses on explaining the "shared interpretive schemes people use to make sense of product innovation," of which two act as barriers to the "development of technology-market knowledge: departmental thought worlds and organizational product routines" (p. 179).

Although inter-departmental collaboration improves the chances of successfully combining a product's technological functionality with market forces, which would result in improvements to both product design and development, Dougherty cites previous research to establish that different people have different "interpretive schemes" that present "major barriers to linking and collaboration." She identifies "two interpretive schemes [that] become interpretive barriers" in the case of innovation: First, departments develop their own "thought worlds," each concentrating on "different aspects of technology-market knowledge," resulting in a specific and most likely different idea of the comprehensive picture. Second, there are routines and patterns of behavior that exacerbate the distinctions and differences among thought worlds. Organizations considering strategies to overcome these impediments to innovation should understand that both cultural and structural solutions are required (p. 179).

Dougherty suggests that understanding the methods and value of successful technology-market linking requires insight into both the linkage's processes and content. Linking demands that new knowledge is developed about both the new product and its intended market. Innovators are adept at reorganizing existing and accepted knowledge and procedures to accommodate new knowledge and requisite alterations to the development process. One challenge to this linkage is understanding, balancing, and choosing "among multiple design options, each with different outcomes" (p. 181).

In terms of content, knowledge from all contributing domains must be collected, particularly from new product developers, as they have "more insight into users' applications, technological trends, distribution systems, and market segments" and can "determine key trends in both the technology and market areas, and then search for 'lead users' who can identify viable design specifications," in similar fashion to the communication stars that Allen and Cohen discuss in Chapter IV (Dougherty, 1992, p. 181).

A major barrier to innovation in large firms is routinization, which has been discussed in technology management literature for at least the past 50 years. Routinization can stymie new product development by limiting any one department's scope of knowledge to that which will fulfill their specific needs. Routines are standard operating procedures that, from a managerial point of view, are beneficial, as they make clear to all departments the roles and limits of each. Not only do routines keep the peace among different departments, they promote rote activities that can be overseen with minimal effort.

Dougherty refers in detail to Fleck's *Genesis and Development of a Scientific Fact,* originally published in 1935. Fleck introduced methods and "insights [helpful to] understanding...the organizational and interpretive processes underlying product innovation and technology-market linking...emphasiz[ing] the social basis of cognition, and adds that innovations often are epistemologically unsolvable by any one person. They require insights from a variety of specialties, called 'thought collectives' or 'thought worlds' as Douglas [in *How Institutions Think*] proposes to retranslate the term. A thought world is a community of persons engaged in a certain domain of activity who have a shared understanding about that activity." Important facets of this concept as it relates to product innovation include the epistemological "fund of knowledge," addressing questions of content; and the metaphysical "systems of meaning," addressing questions of how we know what we know (Dougherty, 1992, p. 182).

Each thought world is shaped by what has already been accepted as knowledge, and different thought worlds hold their storehouse of knowledge, share it among its members, and view other thought worlds' knowledge bases and "central issues as esoteric, if not meaningless." As a thought world evolves, those who participate in one rely on it to shape their perception of new ideas and establish work routines, essential to producing "an 'intrinsic harmony' for the thought world, so ideas that do not fit may be reconfigured or rejected outright." This promotes specialization, which limits and may effectively ignore "information that may be equally essential to the total task" (p. 182).

Dougherty combines Fleck's and Douglas' thought world concepts to other literature concerning routinization, viewing the former's influence on the latter as a negative force for innovation, as it "partition[s] the information and insights" of individual departments. As each thought world necessarily filters information to reinforce a department's understanding of issues, putting barriers up between work groups, it thereby shapes work routines by "providing for only limited interaction, and further inhibit[s] the kind of collective action that is necessary to innovation" (Dougherty, 1992, p. 195).

Information Technology and Competitive Advantage

Sethi and King (1994) operationalized "the construct 'competitive advantage provided by and information technology application (CAPITA)'" so that such an advantage could be quantified, analyzed, and reported to readers interested in finding ways to validate their IT expenditures. "A field study gathered data from 185 top information systems executives regarding information technology applications which had been developed to gain competitive advantage. A confirmatory analysis revealed that CAPITA may be conceptualized in terms of nine dimensions which…form the basis of a preliminary multidimensional measure or index of competitive advantage which has practical uses for competitive advantage" (p. 1601).

As difficult as it has been to assess the degree and nature of the impact that IT investment make, "as increased competition forced firms to scrutinize closely all investments, the issue has become even more critical." Past IT evaluation measures (see King & Epstein, 1983; Bailey & Pearson, 1983; Ives, Olsen, & Baroudi, 1983; Srinivasan, 1985) form the foundation for researchers who "need to assess the strategic role of technology [and] the impact of IT on competitive advantage.…Measurement of competitive advantage (CA) is necessary for choosing between candidate IT applications during the information systems planning process (Geise, 1984; Rackoff, Wiseman, & Ullrich, 1985; Porter & Millar, 1985)" (Sethi & King, 1994, p. 1601). This is because "measures are required to demonstrate and justify the value of IT to top management; currently, two-thirds of Fortune 100 companies' chief executive officers believe that their firms are not getting the most for their IT investments (Rifkin, 1989). Measures of CA are also required in order to conduct empirical studies involving IT effectiveness (Bakos, 1987). In addition, they are essential to further understand concepts such as sustainability and contestable competitive advantage (Ghemawat, 1986).…Thus, in order to move from anecdote and case studies to testable models and hypotheses, it is critical to link theoretical concepts such as CA to empirical indicants" (Sethi & King, 1994, p. 1602).

In order to measure the CAPITA construct, the dimensions were "empirically validated, compared with alternative measures, and if acceptable, subjected to careful replication." Sethi and King's focus is on "the theoretical underpinnings of the CAPITA construct, the substantive issues in its operationalization, and [they use] confirmatory factor analysis to develop the measure." The construct is "rooted in several different concepts, [including] competitive advantage...business value... operational efficiency...management productivity...competitive force...strategic thrust...[and] value activities....These different concepts signify two fundamental approaches to the measurement of competitive advantage. The first, which may be labeled the *outcome approach,* is reflected in concepts such as competitive efficiency, business value, and management productivity...[suggesting that researchers should be] assessing CA using outcomes as the dominant criterion" (p. 1602).

However, the problem with the outcome approach "is that outcome measures at the enterprise level...have many limitations [and] are of little help in understanding 'how' IT affects CA" (Crowston & Treacy, 1986). The "*trait approach* identifies key traits or attributes that characterize competitive advantage. It is reflected in concepts such as competitive forces, strategic thrusts, value activities, and customer resource lifecycle. The trait approach suggests that CA is embodied by the degree to which an IT application possesses certain key attributes. The trait approach is an application of the broader systems resource model, which defines effectiveness as the attainment of a normative state and advocates the measurement of 'means' (Hamilton & Chervany, 1981). The predominant advantage of this approach is that it would provide insights into how and why IT affects CA." Even though the authors studied both approaches, "the trait approach is adopted because the construct development and measurement, the underpinnings of the trait approach, can lay the foundation for theory construction and thus contribute more to the development of the field at the current stage. Nonetheless, the results must be compared with those of the outcome approach in the future" (Sethi & King, 1994, p. 1603).

The authors explain that the first step "in construct operationalization is to delineate its domain. One major factor largely circumscribes the domain of CAPITA: the level of impact of IT [that] is experienced at a number of different organizational levels." Crowston and Treacy (1986) find that "there are three levels for studying IT's impact: internal strategy; competitive strategy[;] and business portfolio strategy." The traits that form the basis for CAPITA "pertain to the impact of an IT application on the competitive position of the organization in the industry." Sethi and King found "five distinct types of benefits from an IT application." The first is efficiency, which "refers to the extent to which an IT application enables a firm to produce products at a lower price relative to competing products. It is conceptualized as resulting from improvements in the firm's structure/processes, leading to a degree of the input/output conversion ratio." Second is functionality, which represents "the extent to which an IT application provides the functionality desired by users....The target of functionality is any internal or external user group. For applications whose target is external, functionality ensures that the application will be adopted and retained, thus enabling the firm to distinguish itself from its competition and to improve its market position. For applications whose target is internal…this trait leads to higher acceptance, utilization, and thus the attainment of intended strategic objectives" (Sethi & King, 1994, p. 1605).

Threat is the third dimension and "refers to the impact of the IT application on the bargaining power of customers and suppliers. It is based on the concept of switching and search-related costs.…Threat enhances the dependence of customers and suppliers on the firm, permitting higher profitability." The fourth dimension, preemptiveness, "characterizes an early and successful preemption of the market by the application. This trait captures the notion of preemptive strikers and leadership technology strategy. Preemptiveness enables a firm to enjoy first-mover advantages, also called 'advantages of occupancy', subsequently unavailable to competitors." Fifth, "synergy refers to the IT application's integration with business goals, strategies, and environment. Such an integration ensures that the application leverages an intrinsic strength of the business, making it difficult for competitors to benefit from copying the application" (pp. 1605-1606).

Sethi and King have determined that "the *efficiency* dimension of CAPITA is not unidimensional as envisioned, but instead comprises two dimensions: Primary Activity Efficiency and Support Activity Efficiency. These results correspond to those of Lind and Zmud (1991), who found that IT support of value chain activities encompasses two components: "impact on primary activities and impact on support activities." Primary activities "involve the 'physical creation of the product and its sale and transfer to the buyer as well as after sale service' (Porter, 1985, p. 18). The results suggest that reducing the cost of activities concerned with the physical creation, distribution, and service of the product is an especially powerful source of CA. On the other hand, reducing the cost of marketing and sales, the remaining primary activity, may directly affect the quality of the offering provided to customers, thus compromising any benefits or competitive advantage resulting from reduced costs" (Sethi & King, 1994, pp. 1613-1614).

"Support Activity Efficiency: Factor 2 [comprises] the impact of the IT application on the cost of the following: human resource management (recruiting, hiring, development and compensation of personnel), firm infrastructure (general management, planning, finance, accounting, legal, government affairs, and quality management), and coordination of different activities" (p. 1614). "Factors 3 and 4 indicate that the hypothesized unidimensional *functionality* trait actually consists of two dimensions: Resource Management Functionality and Resource Acquisition Functionality [which] measures how well the IT application assists its primary users in meeting the following needs related to a resource: monitor utilization, upgrade, transfer or dispose, and account for." The fourth factor, "Resource Acquisition

Functionality, consists of the IT application's impact on the acquisition phase of the resource lifecycle...[measuring] the impact of the IT application on users' ability to order a resource, acquire it, and verify its acceptability." Threat is the fifth factor, consisting "of the impact of the IT application on the following six items: (1) the firm's ability to evaluate and choose from alternate suppliers (supplier selection), (2) switching costs, (3) its ability to threaten vertical integration (both forward and backward) (4) its ability to evaluate and choose alternate customers (customer selection), (5) customers' cost of locating alternate suppliers, [and] (6) customers' switching costs." Preemptiveness is the sixth factor, consisting of "four items: the extent to which the IT application provides unique access to channels (brokers, distributors, and retailers), forces competitors to adopt less favorable market postures, influences the development of industry standards and practices, and offers barriers against imitation such as patents, copyrights, and trade secrets." The final factor, synergy, "is a function of the application's alignment with the firm's business strategy, marketing policies and practices, ability to continuously innovate and enhance the application, technical expertise, and top management support for the application...Synergy represents an exploitation of the firm's uniqueness by the IT application that competitors cannot benefit from or copy" (pp. 1614-1616).

Adaptive Structuration and Organizational Change

DeSanctis and Poole (1994) move the discussion of interaction with computing systems from the individual end user to the firm-wide strategic level by proposing "adaptive structuration theory (AST) as a framework for studying variations in organization change that occur as advanced technologies are used" (p. 122). By refining "structurational concepts to the realm of advanced information technologies," the authors sought to integrate "concepts from the decision-making school with structuration concepts, and demonstrated how structuration can be studied within an empirical program of research" (p. 122).

With the advent and refinement of graphical user interfaces associated with "advanced information technologies, which include electronic messaging systems, executive information systems, collaborative systems, [and] group decision support systems," came "other technologies that use sophisticated information management to enable multiparty participation in organization activities." Although systems developers and end users "hold high hopes for their potential to change organizations for the better, ...actual changes often do not occur, or occur inconsistently." DeSanctis and Poole "propose adaptive structuration theory (AST)...for studying the role of advanced information technologies in organization change [because it] examines the change process from two vantage points: (1) the types of structures that are provided by advanced technologies, and (2) the structures that actually emerge in human action as people interact with these technologies" (p. 121).

Their study, intended to "illustrate the principles of AST...consider[s] the small group meeting and the use of a group decision support system (GDSS)...because it can be structured in a myriad of ways, and social interaction unfolds as the GDSS is used. Both the structure of the technology and the emergent structure of social action can be studied." Their article begins "by positioning AST among competing theoretical perspectives of technology and change...[continues by describing] the theoretical roots and scope of the theory as it is applied to GDSS use and state[s] the essential assumptions, concepts, and propositions of AST...[then outlines] an analytic strategy applying AST principles and provide[s] an illustration of how our analytic approach can shed light on the impacts of advanced technologies on organizations." The authors find that a "major strength of AST is that it expounds the nature of social structures within advanced information technologies and the key interaction processes that figure in their use. By capturing these processes and tracing their impacts, we can reveal the complexity of technology-organization relationships. We can attain a better understanding of how to implement technologies, and we may also be able to develop improved designs or educational programs that promote productive adaptations" (p. 121).

Technologies are often understood to produce "effects [that are] less a function of the technologies themselves than of how they are used by people. For this reason, actual behavior in the context of advanced technologies frequently differs from the 'intended' impacts." Because "people adapt systems to their particular work needs, or they resist them or fail to use them at all...there are wide variances in the patterns of computer use and, consequently, their effects on decision making and other outcomes." Structuration and appropriation, "the central concepts of AST...provide a dynamic picture of the process by which people incorporate advanced technologies into their work practices. According to AST, adaptation of technology structures by organizational actors is a key factor in organizational change" (p. 122).

Decision science maintains a perspective "rooted in the positivist tradition of research and presumes that decision making is 'the primordial organizational act' [emphasizing] the cognitive processes associated with rational decision-making and [adopting] a psychological approach to the study of technology and change." This discipline offers "'systems rationalism'...the view that technology should consist of structures (e.g., data and decision models) designed to overcome human weaknesses (e.g., 'bounded rationality' and 'process losses')" as an alternative to AST. "Once applied, the technology should bring productivity, efficiency, and satisfaction to individuals and organization" (p. 122).

Alternatively, "AST provides a model that describes the interplay between advanced information technologies, social structures, and human interaction. Consistent with structuration theory, AST focuses on social structures, rules and resources provided by technologies and institutions as the basis for human activity," as "advanced information technologies are but one source of structure for groups. The content and constraints of a given work task are another major source of structure" (p. 125). Although "structuration models go beyond the surface of behavior to consider the subtle ways in which technology impacts may unfold," AST does have limitations, which "to date have been their weak consideration of the structural potential of technologies in general and advanced information technologies in particular, their exclusive focus on institutional levels of analysis, and reliance on purely interpretive methods." DeSanctis and Poole suggest that in order to "yield useful knowledge for organizations, structuration-based theories of technology-induced change must devise detailed models of group dynamics and a set of methods for directly investigating the relationship between structure and action" (p. 142).

Situated Practice and Organizational Change

Orlikowski (1996) looks at the relationships between organizational change and information technologies to provide situated practice as a response to "three perspectives that have influenced studies of technology-based organizational transformation—planned change, technological imperative, and punctuated equilibrium....Planned change models presume that managers are the primary source of organizational change, deliberately [initiating and implementing] change in response to perceived opportunities to improve organizational performance." She notes that detractors from planned change models treat "change as a discrete event to be managed separately from the ongoing processes of organizing, and placing undue weight on the rationality of managers directing the change." Second on the list of traditional models of change is the technological imperative perspective, in which "Technology is seen as a primary and relatively autonomous driver of organizational change, so that the adoption of new technology creates predictable changes in organizations' structures, work routines, information flows, and performance." This approach's "deterministic logic...is incompatible with the open-ended nature of many new technologies which assume considerable user customization." Last of the three traditional perspectives is the punctuated equilibrium model, which assumes change is "rapid, episodic, and radical," in which "discontinuities are typically triggered by modifications in environmental or internal conditions [and are] premised on the primacy of organizational stability" (p. 64). The common failing of these three perspectives is their "neglect [of] emergent change. Where deliberate change is the realization of a new pattern of organizing as originally intended, emergent change is the absence of explicit, *a priori* intentions [that] cannot be anticipated or planned" (p. 65).

Organizational change viewed as situated "is premised on the primacy of organizing practices [and] questions the beliefs that organizational change must be planned, that technology is the primary cause of technology-based organizational transformation, and that radical changes always occur rapidly and discontinuously." The situated perspective "is grounded in the ongoing practices of organizational actors, and emerges out of their (tacit and not so tacit) accommodations to and experiments with the everyday contingencies, breakdowns, exceptions, opportunities, and unintended consequences that they encounter.... Organizational transformation is seen here to be an ongoing improvisation enacted by organizational actors trying to make sense of and act coherently in the world" (p. 65).

Situated change "is not an abrupt or discrete event, neither is it, by itself, discontinuous. Rather, through a series of ongoing and situated accommodations, adaptations, and alterations (that draw on previous variations and mediated future ones), sufficient modifications may be enacted over time [so] that fundamental changes are achieved." Rather than considering methods and behaviors as being static, consistent, and predictable, Orlikowski's model of situated change "is grounded in assumptions of action" (p. 66). She observes situated change in organizations that are "implementing and using new information technology," and provides specific, detailed, and convincing evidence of its process in her study of a software company's customer service department's routines—and employee responses to a transition from paper and pencil note-taking to a computer-based information system—to track customer calls. Through her analysis of "work practices, organizing structures, and coordination mechanisms," she developed the "practice-based perspective which offers a conceptual lens with which to focus on types of transformation not discernable to the perspectives of planned change, technological imperative, and punctuated equilibrium. The situated change perspective is offered as a complement to, not a substitute for, the existing change perspectives;" her study "reveals the critical role of situated change enacted by organizational members using the technology over time" (p. 67).

The fundamental finding from her study is that the transformation she finds, "while enabled by the [implementation of] the technology, was not caused by it. Rather, it occurred through the ongoing, gradual, and reciprocal adjustments, accommodations, and improvisations enacted by the CSD [customer support department] members." What is most insightful about her analysis of CSD members is her finding that "their action subtly and significantly altered the organizational practices and structures of the CSD workplace over time, transforming the texture of work, nature of knowledge, patterns of interaction, distribution of work, forms of accountability and control, and mechanisms of coordination." As a result of the CSD members' adoption and use of a call-tracking system imposed by senior management—one that literally moved employees from paper note-taking after a call has been taken to key-stroking customer identification information, the problem's contextual information, and the steps and process that the CSD member took, in real time—Orlikowski posits a structuring process as a "recursive relationship between the everyday actions of human agents and the social structures which are both medium and outcome of those actions." The role of information technology in organizational structuration as defined by DeSanctis and Poole, among others, is refined here to "posit technology not as a physical entity or social construction, but as a set of constraints and enablements realized in practice by the appropriation of technological features." Technology plays "a role similar to that of organizational properties—shaping the production of situated practices, and being shaped by those practices in turn" (p. 69).

The study details what Orlikowski terms five metamorphoses, each characterized by" (i) an analysis of the practices which enacted the changes, including organizational properties which influenced and which were influenced by those changes; (ii) the specific technological features which were appropriated in use; and (iii) the unanticipated outcomes which resulted from the changes and which influenced further changes." Specifically, the five metamorphoses concern organizational changes related to:

1. "The shift to electronic capture, documentation, and searching of call records in [the system's] database";
2. "The redistribution of work from individual to shared responsibility";
3. "The emergence of a proactive form of collaboration among the [call center] specialists";
4. "Expanding into a global support practice, and with creating interdepartmental and cross-functional linkages"; and
5. "Controlling access to and distributing extracts of the knowledge contained within the [system's] database" (pp. 69-70).

Orlikowski's perspective "provides a way of seeing that change may not always be as planned, inevitable, or discontinuous as we imagine. Rather, it is often realized through ongoing variations which emerge frequently, even imperceptibly, in the slippages and improvisations of everyday activities" (pp. 88-89). The metamorphoses described in the study depict changes that "were not planned *a priori,* and neither were they discrete events. Rather, they revealed a pattern of contextualized innovations in practice enacted by all members of the CSD and proceeding over time with no pre-determined endpoint...changes that emerged and evolved through moments of situated practice over time"—limiting the power that theories of punctuated equilibrium have in predicting the nature of organizational change (p. 89).

When tools "enable users to construct and customize specific versions and local adaptations of [their] underlying technological features...two important implications for practice" result. The first is that it "allows for easy and ongoing changes to the technology in use, in contrast to more rigid, fixed-function technologies which are difficult and costly, if not impossible, to change during use. Two, because customization is required for effective use, technological and organizational changes are encouraged." A "perspective of situated change, which by definition, assumes context specificity," combines technological functions, organizational structures, and human agency to describe the process of "ongoing local improvisations in response to deliberate and emergent variations in practice," what the author deems as "potentially generalizable and is [therefore] offered as a stimulus for further research" (p. 90).

TAM Revisited

Venkatesh and Davis (2000) extend Davis's Technology Acceptance Model by considering additional variables and their effects on perceived usefulness and perceived ease of use of information technologies. Their longitudinal studies consist of following two firms whose employees were in a position to voluntarily adopt new software and platforms, and two for whom technology adoption was mandatory. Their study was instigated by the fact that "information technology adoption and use in the workplace remains a central concern for information systems research and practice. Despite impressive advances in hardware and software capabilities, the troubling problem of underutilized systems continues" (p. 186). Given the general acceptance of TAM as a reliable theoretical construct, one whose relevance is grounded in the fact that perceived usefulness, "defined as the extent to which a person believes that using the system will enhance his or her job performance, and the perceived ease of use, defined as the extent to which a person believes that using the system will be free of effort," the authors hypothesize that expanding TAM to incorporate "additional theoretical constructs spanning social influence (subjective norm, voluntariness, and image) and cognitive instrumental processes (job relevance, output quality, result demonstrability, and perceived ease of use)" will both strengthen the validity of TAM and respond favorably to the increasing demands and opportunities that information technologies and systems place upon and offer organizations.

The authors begin by defining and describing the effects of TAM2's "impacts [on] three interrelated social forces impinging on an individual facing the opportunity to adopt or reject a new system: subjective norm, voluntariness, and image" (p. 187). Subjective norm is "a person's perception that most people who are important to him think he should or should not perform the behavior in question," and derives from the rationale that "people may choose to perform a behavior, even if they are not themselves favorable toward the behavior or its consequences, if they believe one or more important referents think they should, and they are sufficiently motivated to comply with the referents." Venkatesh and Davis expand upon Davis et al.'s (1989) original TAM model by heeding his suggestion deriving from the finding that "subjective norm had no significant effect on intentions over and above perceived usefulness and ease of use...but they did acknowledge the need for additional research to 'investigate the conditions and mechanisms governing the impact of social influences on usage behavior' [999]" (Venkatesh & Davis, 2000, p. 187).

The literature on voluntariness and compliance in relation to subjective norm finds that "subjective norm had a significant effect on intention in mandatory settings but not in voluntary settings. We refer to the causal mechanism underlying this effect as compliance. In general, the direct compliance effect of subjective norm on intention is theorized to operate whenever an individual perceives that a social actor wants him or her to perform a specific behavior, and the social actor has the ability to reward the behavior or punish nonbehavior. TAM2 theorizes that, in a computer usage context, the direct compliance-based effect of subjective norm on intention over and above perceived usefulness and perceived ease of use will occur in mandatory, but not voluntary, system usage settings....[TAM2] posits voluntariness as a moderating variable, defined as the extent to which potential adopters perceive the adoption decision to be non-mandatory" (p. 188).

TAM2 moves beyond the Theory of Reasoned Action and the Theory of Planned Behavior by taking into account that "subjective norm can influence intention indirectly through perceived usefulness, [incorporating] internalization and identification. Internalization refers to the process by which, when one perceives that an important referent thinks one should use a system, one incorporates the referent's belief into one's own belief structure....Social influence is defined as influence to accept information from another as evidence about reality." TAM2 proposes that "internalization, unlike compliance, will occur whether the context of system use is voluntary or mandatory. That is, even when system use is organizationally mandated, users' perceptions about usefulness may still increase in response to persuasive social information" (p. 189).

In terms of reactions to image, with image defined as "the degree to which use of an innovation is perceived to enhance one's...status in one's social system...TAM2 theorizes that subjective norm will positively influence image because, if important members of a person's social group at work believe that he or she should perform a behavior, then performing it will tend to elevate his or own standing within the group." Identification effect refers to "the effect of subjective norm on image, coupled with the effect of image on perceived usefulness. TAM2 theorizes that identification, like internalization but unlike compliance, will occur whether the context of system use is voluntary or mandatory" (p. 189).

The authors' longitudinal studies also examined how experience affects changes in social influence. "[Before] a system is developed, users' knowledge and beliefs about a system are 'vague and ill-formed', and they must therefore rely more on the opinions of others as a basis for their intentions (Hartwick & Barki, 1994, pp. 458-459). After implementation, when more about the system's strengths and weaknesses are known through direct experience, the normative influence subsides....TAM2 theorizes that the direct effects of subjective norm on the intentions of mandatory usage contexts will be stronger prior to implementation and during early usage, but will weaken over time as increasing direct experience with a system provides a growing basis for intentions toward ongoing use." Venkatesh and Davis (2000) also expect "the effect of subjective norm on perceived usefulness (internalization) to weaken over time, since greater direct experience will furnish concrete sensory information, supplanting reliance on social cues as a basis for usefulness perceptions." However, this hypothesis runs contrary to their expectations that the influence of image on perceived usefulness (identification) will "weaken over time since status gains from system use will continue as long as group norms continue to favor usage of the target system" (p. 190).

Equally important to this article, as social influences are, is Venkatesh and Davis's discussion of cognitive instrumental processes and the complexities they add to the expansion of TAM to TAM2. The authors "theorize four cognitive instrumental determinants of perceived usefulness: job relevance, output quality, result demonstrability, and perceived ease of use...people form perceived usefulness judgments in part by cognitively comparing what a system is capable of doing with what they need to get done in their job" (p. 190). "TAM2 theorizes that people use mental representation for assessing the match between important work goals and the consequences of performing the act of using a system as a basis for forming judgments about the use-performance contingency (i.e., perceived usefulness)" (p. 191).

Job relevance is defined as "an individual's perceptions regarding the degree to which the target system is applicable to his or her job." It is "a function of the importance within one's job of the set of tasks the system is capable of supporting." Notwithstanding what a system can do and the level of congruence between a system and someone's job tasks, TAM2 "posits that...people will take into consideration how well the system performs those tasks, which we refer to as perceptions of *output quality*" (p. 191). Result demonstrability, or the "tangibility of the results of using the innovation, will directly influence perceived usefulness. This implies that individuals can be expected to form more positive perceptions of the usefulness of a system if the covariation between usage and positive results is readily discernable. Conversely, if a system produces effective job-relevant results desired by a user, but does so in an obscure fashion, users of the system are unlikely to understand how useful such a system really is" (p. 191). The authors hypothesize that the combination of social influence and cognitive instrumental processes will contribute to understanding why there is "a decrease in the strength with which social influence processes affect perceived usefulness and intention to use with increasing experience over time" (p. 192).

The authors' hypotheses were proven out by the data provided by the four studies. "Based on measures of voluntariness...we confirmed that employees in Studies 1 and 2 perceived system use to be voluntary...whereas employees in Studies 3 and 4 perceived system use to be mandatory." The data presented show that "perceived usefulness was a strong determinant of intention to use, and perceived ease of use was a significant secondary determinant. The effect of subjective norm on intention (compliance) was consistent with our expectations." These data provide a contrast, however, "in cases where usage was voluntary, [whereas] subjective norm had no direct effect on intention over and above what was explained by perceived usefulness and perceived ease of use. Unlike subjective norm, perceived usefulness and perceived ease of use remained consistent significant determinants of intention across all time periods in all four studies." Intention "fully mediated the effects of perceived usefulness, perceived ease of use, and subjective norm on usage behavior" (p. 195).

In each of the four studies, subjects were surveyed three times: once before the new system was implemented, once one month into the implementation, and once three months after system implementation. "TAM2 explained up to 60% of the variance in perceived usefulness. The effect of subjective norm on perceived usefulness (internalization) was significant at T1 [time 1] and T2 [time 2] but weakened by T3 [time 3]. The influence of image on perceived usefulness (identification) was significant at all three points of measurement. Also, as hypothesized, the effect of subjective norm on image was significant at all points of measurement....the interaction of job relevance and output quality was significant in all four studies at all points of measurement" (p. 196). In sum, "perceived usefulness fully mediated the effects of all of its determinants on usage intentions, except for the expected direct effects of perceived ease of use in all studies and subjective norm for the mandatory usage studies (3 and 4)" (p. 197).

As the data show, the effects of both social influence and cognitive instrumental processes were consistent with TAM2. "An important and interesting finding that emerged was the interactive effect between job relevance and output quality in determining perceived usefulness. This implies that judgments about a system's usefulness are affected by an individual's cognitive matching of their job goals with the consequences of system use (job relevance), and that output quality takes on greater importance in proportion to a system's job relevance" (p. 199). Given all the data, the authors derive "several practical implications," including "mandatory, compliance-based approaches to introducing new systems appear to be less effective over time than the use of social influence to target positive changes in perceived usefulness...in addition to designing systems to better match job output, or making them easier to use...practical interventions for increasing results demonstrably, such as empirically demonstrating to users the comparative effectiveness of a new system relative to the status quo may provide important leverage for increasing user acceptance" (p. 199). The authors conclude by indicating that the "continuing trend in organizations away from hierarchical, command-and-control structures toward networks of empowered autonomous teams underscores our finding regarding the limits of organizational mandate as a lever for increasing usage. To adapt user acceptance theory to this trend, the conceptualization of perceived usefulness will need to be expanded from its current focus on expected individual-level performance gains to encompass team-based structures and incentives" (p. 200).

CONCLUSION

Much has happened to the field of Communication and Information Technology, both in terms of theory and practice, since McFarlan (1984) recognized the potential of the Internet and other technologies as being critical to strategy and competition, due to their potential to change the nature of competition by reducing and establishing new barriers to entry. How information technologies can supply competitive advantage to firms is the subject of Sethi and King (1994). To assist senior managers assess the current and potential impact of IT on their businesses, a multilevel framework for assessing the competitive impact of information technology on a firm was developed (Parsons, 1983) and consists of:

1. **Industry-level impacts:** Products and services, markets, and production economics;
2. **Firm-level impacts—buyers:** Suppliers, substitution, new entrants, and rivalry; and
3. **Strategy-level impacts:** Low-cost leadership, product differentiation, and concentration on market or product.

While some of the consideration of information technology has focused on how to recognize and manage these technologies for obtaining strategic advantage, a great deal of consideration has been given to information technology's role in changing the processes used by an organization, and some attention has also been given to the development of information technologies as a product. The role of information technologies in computer-assisted collaborative workspaces is the subject of Stefik et al.'s (1987) work at Xerox PARC's original Colab environment.

One of the ways that information technology is differentiated from other types of technological innovation is the focus that has been given to the activities that occur after the purchase decision is made by the firm. This is in part the role information technology has in visibly transforming the processes and often structure throughout a firm. Orlikowski (1996) notes that these changes or metamorphoses are due to:

1. The move to electronic capture, documentation, and searching for records in databases;
2. Controlling access to and distributing extracts of knowledge contained within databases;
3. Emergence of a proactive form of collaboration among specialists;
4. Redistribution of work from individual to shared responsibility; and
5. Expanding into a global support practice with the creation of interdepartmental and cross-functional linkages.

Adaptive Structuration Theory (AST) helps to address these concerns because it offers a framework for studying variations in organization change that occur as advanced technologies are used. AST examines the:

1. Types of structures that are provided by advanced technologies, and
2. Structures that actually emerge in human action as people interact with these technologies (DeSanctis & Poole, 1994).

The insights that this theory offers assist in obtaining a better understanding of how to implement technologies, improved designs, and develop educational programs that promote effective adaptations.

The development of information technology is considered by Dougherty (1992), who finds that:

1. Product success is positively correlated to the closeness in fit between product design and customer need.
2. Collaboration between departments in a large corporation increases the chances of the product's success.
3. Innovators often work without the benefit of either inter-departmental collaboration or linking the technological and market issues involved in their product.

Having considered the frequently cited development literature that is specific to information technology, attention is now turned to the process of integrating information technology into an organization. The fundamental problem with MIS is system designers' not factoring in users' psychological and behavior responses to a system, and/or the organizational culture in which the MIS is to be used. This leads to an expectancy-based model of user reaction and job motivation (Robey, 1979). Individual differences most relevant to MIS success can be placed into three classes of variables: cognitive style, personality, and demographic/situational (Zmud, 1979).

Through the consideration of MIS user pre-implementation expectations, Ginsberg (1981) finds that a great deal is riding on the proper management of user expectations. To reduce failure, attention should be placed on disconfirming expectations of performance satisfaction, product evaluation, and job satisfaction. If users have a realistic sense of what they can expect from a management information system before the design is implemented, the project outcome, in terms of users' attitudes and behaviors, will be positive. The constructs of user involvement and user satisfaction have been found to be critical to integration of information technology into an organization, consequently they are considered next.

User involvement is generally understood to be a necessary component of evaluating information systems due to its perceived relationship with system success. However, a review of the relevant organizational behavior literature found that there is only a weak connection between established theory and the premises that researchers use in designing their research questions on user involvement (Ives & Olson, 1984), raising the question of the presence of the link between user involvement and system success. The user satisfaction construct is critical to MIS research. Similarities have been found between user satisfaction and behavioralist definitions of attitude (Melone, 1990). When measuring effectiveness, user satisfaction alone is insufficient to adequately capture the full meaning of effectiveness. Output-oriented criteria should also be utilized, as this includes the cognitive and behavioral components of attitude. It is critical to note that attitude is much more complex than is often conceptualized in user-satisfaction research.

The Theory of Reasoned Action (TRA) and Technology Acceptance Model (TAM) are used to predict and explain user acceptance and rejection of computer-based technology (Davis et al., 1989). More specifically, Davis et al. inquire into how well TAM can predict and explain future user behavior from simple measures taken after a very brief period of interaction with a system. TAM has been found to have value in determining user acceptance, with practical value for evaluating systems and guiding managerial interventions to reduce the occurrence of underutilized computer technology. Because TAM is seen as a reliable theoretical construct, one whose relevance is grounded in observations of perceived usefulness, Venkatesh and Davis (2000) expand TAM to incorporate additional theoretical constructs spanning social influence and cognitive instrumental processes. Thereby, they strengthen the validity of TAM and respond to increasing demands and opportunities that information technologies and systems place upon and offer organizations. Through the examination of information technology's role in adding value and changing the nature of competition, organizational processes, and structure, it can be seen why information technology is considered separately from other classes of innovation. Having addressed these differences, open source and software development innovation are considered.

REFERENCES

Ajzen, I., & Fishbein, M. (1980). *Understanding attitudes and predicting social behavior.* Englewood Cliffs, NJ: Prentice Hall.

Bailey, J., & Pearson, S. (1983). Development of a tool for measuring and analyzing computer user satisfaction. *Management Science, 29*(5), 530-545.

Bakos, Y. (1987, December). Dependent variables for the study of firm and industry-level impacts of information technology. *Proceedings of the 8th International Conference on Information Systems* (pp. 10-23), Pittsburgh, PA.

Crowston, K., & Treacy, M. (1986, December). Assessing the impact of information technology on enterprise level performance. *Proceedings of the 7th Annual International Conference on Information Systems* (pp. 299-310), San Diego, CA.

Davis, F. (1986). *A technology acceptance model for empirically testing new end-user information systems: Theory and results.* Doctoral Dissertation, Sloan School of Management, Massachusetts Institute of Technology, USA.

Davis, F., Bagozzi, R., & Warshaw, P. (1989). User acceptance of computer technology—a comparison of two theoretical models. *Management Science, 35*(8), 982-1003.

Davis, G.B. (1974). *Management information systems: Conceptual foundations, structure and development.* New York: McGraw-Hill.

DeSanctis, G., & Poole, M.S. (1994). Capturing the complexity in advanced technology use—an adaptive structuration theory. *Organization Science, 5*(2), 121-147.

Dougherty, D. (1992). Interpretive barriers to successful product innovation in large firms. *Organization Science, 3*(2), 179-202.

Douglas, M. (1987). *How institutions think.* London: Routledge and Kegan Paul.

Fishbein, M., & Ajzen, I. (1975). *Belief, attitude, intention and behavior: An introduction to theory and research.* Reading, MA: Addison-Wesley.

Fleck, L. (1979). *Genesis and development of scientific fact* (translated by F. Bradley & T. Trenn; originally published 1935). Chicago: University of Chicago Press.

Geise, P. (1984). Using information technology to capture strategic position. *Management Review, 73*(9), 8-11.

Ghemawat, P. (1986). Sustainable advantage. *Harvard Business Review, 64*(5), 53-58.

Ginzberg, M.J. (1981). Early diagnosis of MIS implementation failure: Promising results and unanswered questions. *Management Science, 27*(4), 459-478.

Hamilton, S., & Chervany, N. (1981). Evaluating information system effectiveness—part I: Comparing evaluation approaches. *MIS Quarterly, 5*(3), 55-69.

Hartwick, J., & Barki, H. (1994). Explaining the role of user participation in information systems use. *Management Science, 40,* 440-465.

Ives, B., & Olson, M. (1984). User involvement and MIS success: A review of research. *Management Science, 30*(5), 586-603.

Ives, B., Olsen, M., & Baroudi, J. (1983). The measurement of user information satisfaction. *Communications of the ACM, 26*(10), 785-793.

King, W., & Epstein, B. (1983). Assessing information system value: An experimental study. *Decision Science, 14,* 34-45.

Lind, M., & Zmud, R.W. (1991). The influence of a convergence in understanding between technology providers and users on information technology innovativeness. *Organization Science, 2*(2), 195-217.

McFarlan, F.W. (1984). Information technology changes the way you compete. *Harvard Business Review, 63*(3), 98-103.

Melone, N.P. (1990). A theoretical assessment of the user satisfaction construct information systems research. *Management Science, 36*(1), 76-91.

Orlikowski, W.J. (1996). Improvising organizational transformation over time: A situated change perspective. *Information Systems Research, 7*(1), 63-92.

Parsons, G.L. (1983). Information technology—a new competitive weapon. *Sloan Management Review, 25*(1), 3-14.

Porter, M. (1985). *Competitive advantage.* New York: The Free Press.

Porter, M., & Millar, V. (1985). How information gives you competitive advantage. *Harvard Business Review,* (July-August), 149-160.

Rackoff, N., Wiseman, C., & Ullrich, W. (1985). Information systems for competitive advantage: Implementation of a planning process. *MIS Quarterly, 9*(4), 285-294.

Rifkin, G. (1989). CEOs give credit for today but expect more for tomorrow. *Computerworld,* (April 17), 75-82.

Robey, D. (1979). User attitudes and management information system use. *Academy of Management Journal, 22*(3), 527-538.

Sethi, V., & King, W. (1994). Development of measures to assess the extent to which an information technology application provides competitive advantage. *Management Science, 40*(12), 1601-1627.

Srinivasan, A. (1985). Alternative measures of system effectiveness: Associations and implications. *MIS Quarterly, 9*(3), 243-253.

Stefik, M., Foster, G., Bobrow, D., Kahn, K., Lanning, S., & Suchman, L. (1987). Beyond the chalkboard: Computer support for collaboration and problem solving in meetings. *Communications of the ACM, 30*(1), 32-47.

Venkatesh, V., & Davis, F. (2000). Theoretical extension of the Technology Acceptance Model: Four longitudinal field studies. *Management Science, 46*(2), 186-204.

Zmud, R.W. (1979). Individual differences and MIS success—review of the empirical literature. *Management Science, 25*(10), 966-979.

Chapter X
Open Source and Software Development Innovation

INTRODUCTION

It is beyond question how ubiquitous and powerful computing has become for commerce, communication, and culture. As the articles addressed in this chapter make clear, the development of software poses challenges to those with commercial concerns—those that build software and those that use it—as well as specific situations in which management and innovation theory is responsive also to non-proprietary software development. We begin with two articles by Boehm, arguably the most prominent voice in software engineering today. The first, with Ross (1989), introduces advances in theory to aid software project management, and the second (1991) takes a close look at risk management as it pertains to software development projects. Fichman and Kemerer (1997) present their research findings related to knowledge management in software process innovation management environments, while Nambisan and Wilemon (2000) explain the mutually advantageous bodies of knowledge that the realms of software development and new product development hold for one another. Fajar and Sproull (2000) consider software development management from a knowledge and team management perspective, and their findings have affinities with Farris et al.'s (2003) introduction of the Web of Innovation, which facilitates an organization's e-knowledge management systems and their application to new product development.

The growth of the open source software movement is the subject of this chapter's five concluding articles. While we suggest readers take note of E. Raymond's *The*

Cathedral and the Bazaar (1999) for a thorough and authoritative recounting of the players and development of this highly prolific and out-of-the-ordinary approach to software development, the articles addressed in the second part of this chapter have particular relevance to software's place in the wider scope of technology innovation management. Lerner and Tirole (2002) ask and answer, from an economic perspective, a question that those unfamiliar with software development's history or scope of work might well pose: why would a talented computer programmer give her time, effort, and knowledge away for free? Their answers are relevant to von Hippel and von Krogh's (2003) article on the open source movement's reshaping of the private investment and the collective action models of innovation. Lakhani and von Hippel (2003) dig a little deeper into the rationale of programmers who invest themselves in attending to what some may view as the tedious tasks of software development, and join the others in seeking clues as to how open source commands the time and attention of so many skilled programmers. von Krogh, Spaeth, and Lakhani (2003) develop four theoretical constructs that can assist proprietary firms engaged in software development projects understand how to recruit and maintain a cohort of seasoned programmers by mirroring the mentoring culture that proliferates through the open source movement. The chapter concludes with an overview of West's (2003) article offering sound research directions, based on the experience of IBM, Sun Microsystems, and Apple, for organizations seeking to meld proprietary and open source methodologies and management styles with established theories of appropriability and adoption as they pertain to software development.

Management Theory and Software Development

Boehm and Ross (1989) present a new project management theory for a relatively new industry, software development, in response to what they describe as a proliferation of "various alphabetical management theories": Theory W: Make Everyone a Winner. Theory W has two underlying principles: "plan the flight and fly the plan," and "identify and manage your risks." Boehm and Ross's project in this article is to situate their theory in what was then, and is even more so today, an expansive, intensive, and critical industry, software development, but address this industry from a project manager's point of view. "The skillful integration of software technology, economics, and human relations in the specific context of a software project is not an easy task. The software project is a highly people-intensive effort that spans a very lengthy period, with fundamental implications on the work and performance of many different classes of people" (p. 902).

The project manager is responsible to several different constituencies, some of whom have different, and at times, competing agendas. Given the mix of goals, perspectives, desires, and timetables, there are bound to be "project management difficulties—both at the strategic...and tactical level....A good software management theory should help the project manager navigate through these difficulties [by being] simple to understand and apply; general enough to cover all classes of projects and classes of concerns (procedural, technical, economic, people-oriented); yet specific enough to provide useful, situation-specific advice" (p. 902).

In order to satisfy all of the different parties involved in a software development project, Theory W is proposed to help managers identify and address three different situations that can occur during a project's lifespan: win-win, win-lose, and lose-lose. The goal of Theory W is to identify all the steps a manager will take to produce an outcome in which everyone considers him- or herself a winner. Boehm and Ross compare their Theory W with the scientific management approach introduced by Frederick Taylor—what the authors refer to as Theory X. By studying time and motion in the manufacturing process, managers would "organize jobs into well-orchestrated sequences of tasks in which people were as efficient and predictable as machines." Management's role was to keep "the system running smoothly, largely through coercion" (p. 903).

For an industry that touts its production forces as being responsive to individuals' "creativity, adaptiveness and self-esteem," Theory X is deemed untenable, and as will be clear in the sections on open source, antithetical. In comes Theory Y, which holds that "management should stimulate creativity and individual initiative" (p. 903). The problem with Theory Y, however, is that project development can stumble or be hindered by the conflicts that N number of individuals exercising their creativity can bring about, which introduces the need for Theory Z, holding that "much of the conflict resolution problem can be eliminated by up-front investment in developing shared values and arriving at major decisions by consensus" (p. 904).

Theory W "holds that software project managers will be fully successful if and only if they make winners of all of the participants in the software process...Rather than characterizing a manager as an autocrat (Theory X), a coach (Theory Y) or a facilitator (Theory Z), Theory W characterizes a manager's primary role as a negotiator between his various constituencies, and a packager of project solutions with win conditions for all parties. Beyond this, the manager is also a goal-setter, a monitor of progress toward goals, and an activist in seeking out day-to-day win-lose or lose-lose project conflicts, confronting them, and changing them into win-win situations" (p. 904).

Boehm and Ross borrow and build upon Fisher and Ury's (1983) four negotiation steps:

1. Separate the people from the problem;
2. Focus on interests, not positions;

3. Invent options for mutual gain; and
4. Insist on using objective criteria.

The steps necessary for win-win software project management are:

1. Establish a set of win-win preconditions by understanding how people want to win, establishing reasonable expectations, matching people's tasks to their win conditions, and providing a supportive environment;
2. Structure a win-win software process by establishing a realistic process plan, using the plan to control the project, identifying and managing your win-lose or lose-lose risks, and keeping people involved; and
3. Structure a win-win software project by matching product to users' and maintainers' win conditions (Boehm & Ross, 1989, p. 904).

Risk Management and Software Development

In "Software Risk Management: Principles and Practices," Boehm (1991) provides a primer for software project managers, who increasingly experience high-risk elements that can delay or contribute to failure of their projects. Part of the problem derives from two standard, routinely employed software project development process models: "The sequential, document-driven waterfall process model [that] tempts people to overpromise software capabilities in contractually binding requirement specifications before they understand their risk implications; [and] the code-driven, evolutionary development process model [that] tempts people to say, 'Here are some neat ideas I'd like to put into the system. I'll code them up, and if they don't fit other people's ideas, we'll just evolve things until they work'" (pp. 32-33).

No matter which model or variant of a process model was employed by the project managers that Boehm studied for this article, "One pattern that emerged very strongly was that the successful project managers were good risk managers...they were using a general concept of risk exposure (potential loss times the probability of loss) to guide their priorities and actions." The fact that any software development project by its very nature is composed of different and often competing constituent goals, "it is clear that 'unsatisfactory outcome' is multidimensional: for customers and developers, budget overruns and schedule slips are unsatisfactory. For users, products with the wrong functionality, user-interface shortfalls, performance shortfalls, or reliability shortfalls are unsatisfactory. For maintainers, poor-quality software is unsatisfactory" (p. 33).

To obviate risk, Boehm suggests that software development project managers employ decision trees, constructed either by the project team or outside consultants. In either case, using a decision tree to assess and control risk should be undertaken to identify "critical errors existing or being found and eliminated, their probabilities, the losses associated with each outcome, and the total risk exposure associated" with the project. Decision trees also provide "a framework for analyzing the sensitivity of preferred solutions to the risk-exposure parameters" (p. 34). An associated and equally essential tool is a risk management tree, which comprises "two primary steps each with three subsidiary steps." Step 1, "risk assessment, involves risk identification, risk analysis and risk prioritization. Risk identification produces lists of the project-specific risk items likely to compromise a project's success....Risk analysis assesses the loss probability and loss magnitude for each identified risk item, and it assesses compound risks in risk-item interactions....Risk prioritization produces a ranked ordering of the risk items identified and analyzed....The second primary step, risk control, involves risk-management planning, risk resolution, and risk monitoring. Risk-management planning helps prepare you to address each risk item...including the coordination of the individual risk-item plans with each other and with the overall project plan....Risk resolution produces a situation in which the risk items are eliminated or otherwise resolved...Risk monitoring involves tracking the project's progress toward resolving its risk items and taking corrective action where appropriate" (p. 34).

Boehm's extensive experience with large software development projects is grist for his top-10 list of risk items and individual risk-management techniques to address each item. "Personnel shortfalls" can be addressed in several ways: by "staffing with top talent, job matching, team building, key personnel agreements, and cross training." "Unrealistic schedules and budgets" can be revised with "detailed multisource cost and scheduling estimations, design to cost, incremental development, software reuse, and requirements scrubbing....Developing the wrong functions and properties" can be forestalled through "organization analysis, mission analysis, operations-concept formulation, user surveys and user participation, prototyping, early users' manuals, off-nominal performance analysis, and quality-factor analysis.... Developing the wrong user interface" can be avoided by "prototyping, scenarios, task analysis and user participation." "Gold-plating" can be prevented by "requirements scrubbing, prototyping, cost-benefit analysis, and designing to cost....Continuing stream of requirements changes" can be eluded through a "high change threshold, information hiding, and incremental development (deferring change to later increments)....Shortfalls in externally furnished components" can be avoided through "benchmarking, inspections, reference checking, and compatibility analysis.... Shortfalls in externally performed tasks" can be addressed by "reference-checking, pre-award audits, award-fee contracts, competitive design or prototyping, and team-building....Real-time performance shortfalls" can be averted through "simulation, benchmarking, modeling, prototyping, instrumentation and tuning....Straining computer-science capabilities" can be circumvented through "technical analysis, cost-benefit analysis, prototyping and reference checking" (p. 35).

Ultimately, Boehm identifies a project manager's ability to get focused on critical success factors to be the most important skill. "For various reasons...projects get focused on activities that are not critical for their success.... In the process, critical success factors get neglected, the project fails, and nobody wins. The key contribution of software risk management is to create this focus on critical success factors—and to provide the techniques that let the project deal with them" (p. 40).

Managing Software Process Innovation

Fichman and Kemerer (1997) expand the notion of identifying critical success factors at the managerial level as part of the larger organizational view, particularly organizations that are involved with software process innovations (SPIs). They "seek to explain differences in the propensity of firms to initiate and sustain the assimilation of complex process technologies" by building on Attewell (1992), who sees that "the burden of organizational learning surrounding complex technologies creates a 'knowledge barrier' that inhibits diffusion." A shortfall of the Attewell approach is that "it says little about which organizations should be among the early adopters even in the face of high knowledge barriers." Fichman and Kemerer (1997) seek validation of their hypotheses that "organizations will have a greater propensity to innovate with complex technologies when the burden of organizational learning is effectively lower, either because the organizations already possess much of the know-how and technical knowledge necessary to innovate, or because they can acquire such knowledge more easily or more economically" (pp. 1345-1346).

Specifically, organizations' information technology departments will be more apt to "initiate and sustain the assimilation of software process innovations (SPIs) when they have a greater scale of activities over which learning costs can be spread (*learning-related scale*), more extensive existing knowledge in areas related to the focal innovation (*related knowledge*), and a greater diversity of technical knowledge and activities in general (*diversity*) (p. 1346). To test their hypotheses, Fichman and Kemerer reviewed data collected from over 600 firms whose IT departments assimilated object-oriented programming languages (OOPLs) to develop new software tools. The authors adopt Meyer and Goes' (1988) definition of assimilation, which is "the process spanning from an organization's first awareness of an innovation to, potentially, acquisition and widespread deployment." At the time of their writing, OOPLs were as innovative as CASE tools, relational databases, and fourth-generation programming languages. The authors consider OOPLs innovative "because their deployment results in a change in the organizational processes used to develop software applications," and because of their being subject to knowledge barriers, they "constitute an ideal setting for testing the proposed organizational learning-based model of technology assimilation" (Fichman & Kemerer, 1997, p. 1346).

Where Attewell's approach is deficient, in Fichman and Kemerer's estimation, is that it "leaves open the important question of which end-user organizations can be expected to be among the early adopters, even in the face of high knowledge barriers." Fichman and Kemerer's work seeks to validate their proposal that there "will be those organizations that are less subject to the adverse effects of knowledge barriers, for one or more of three general reasons: (1) they are better positioned…to amortize the costs of learning[,] (2) they have the ability to acquire any given amount of new knowledge with less effort[, and] (3) they will have less to learn about innovation to begin with." Where SPIs are concerned, assimilation involves different kinds of learning. Organizations must understand "the abstract principles on which the technology is based…the nature of the benefits" the technology will afford and the "specific technical features of different commercially available instances of the technology," knowing to which "kinds of problems the technology is best applied…acquiring individual skills and knowledge needed to produce a sound technical product…and designing appropriate organizational changes in terms of team structure, hiring, training and incentives" (Fichman & Kemerer, 1997, p. 1347).

Two very important principles in software development with OOPLs are encapsulation and inheritance, for they "promote such long-standing software engineering goals as abstraction, modularity, reuse, maintainability, extensibility, and interoperability." Such values contribute to the view that "object technology has the potential to bring about an 'industrial revolution' in the way software is constructed, by transforming the current custom development process into one characterized by assembly of previously developed components" (pp. 1348-1349). In the decade following their pronouncement, C++ and other OOPLs have proven to yield these benefits; it is just these capabilities, in conjunction with the proliferation and ease of Internet communication, that have contributed to another production revolution, the open source movement.

Unlike Downs and Mohr (1976) [see Chapter III], whose work is deemed too abstract and theoretical to hold value for particular innovations and environments, Fichman and Kemerer (1997) suggest that researchers "develop theories focusing on the distinctive qualities of particular kinds of technologies…[ones that show] a greater propensity to initiate and sustain the assimilation of SPIs [and how they] will be exhibited by organizations with higher learning-related scale, related knowledge and diversity" (p. 1349). Such research, resulting in a profile of organizations with the likelihood to benefit from SPIs, has practical implications "for technology vendors and mediating institutions, as well as end-users." Organizations that "have more extensive knowledge related to OOPLs…and greater technical diversity in their applications development area" should ramp up their "targeted marketing and promotion...based on how well [specific targets] fit the profile of an early assimilator....Vendors and mediating institutions should...be more focused on identifying appropriate adoption candidates, learning about the particular challenges these organizations face, and taking a more proactive role to promote successful assimilation among these sorts of organizations." As for end user organizations, this research yields the implication that "they would be well advised to view SPI assimilation as a multiyear process of organizational learning, and to carefully assess the extent to which they fit the profile of an early and sustained assimilator" (p. 1360).

Software and New Product Development: Intersections

The potential benefits derived from interdisciplinary sharing of knowledge and theories, particularly when the domains are technology innovation and software development, is the subject of Nambisan and Wilemon (2000). Despite the fact that new product development and software development obviously share some generic complementarities, "Software development (SD) literature emphasizes development methodologies, techniques and process metrics, while new product development (NPD) studies typically focus on organizational factors like teamwork, cross-functional integration, internal/external communication in teams, performance, processes, and project leadership." Nambisan and Wilemon seek to establish "the potential for SD and NPD to learn from each other by addressing" common challenges to both domains, and by "deriving significant cross-domain lessons in three areas, namely, teamwork management, development process maturity, and development process acceleration." Generally, Nambisan and Wilemon view "product development along three dimensions—*people, process, and technology.*" They contend that NPD can teach SD about managing individual people and teams, while SD can demonstrate to NPD its expertise in accelerating processes and maximizing technology (p. 211).

By "technology," Nambisan and Wilemon mean "the different tools and techniques employed in the [product] development....The process dimension includes specific methodologies, project activity planning, scheduling, process standards and metrics, resource management, and continuous learning. The people dimension involves all personnel deployment and management issues" (pp. 212-213). Review of the literature reveals that each domain focuses "on different sets of these three dimensions. Software development scholars have emphasized structured processes and the use of technologies to speed up these processes, i.e., interactions along the process/technology dimensions.....However, applications of software engineering technologies have only produced modest gains in project outcomes....SD research also needs to focus on the people/process dimensions." As for NPD, the focus has been on "the impact of people and their management of the development process, i.e., interactions along the people/process dimensions." The authors speak to the lack of "engineering rigor" in NPD, suggesting that "NPD may have important lessons to learn with respect to the process/technology dimensions" (p. 213).

Improvements in teamwork management can lead to "the early identification of problems, [providing] a coherent customer focus for the development effort," as well as enhancement of "continuous innovation and learning." The more efficient a team is, the more demand there is on "organizations to make specific accommodations to create desirable organizational contexts." NPD research on team formation indicates that "when teams are self-selected...higher commitment and more effective team building is likely to be achieved." Nambisan and Wilemon point out that in the SD domain, "team members are chosen by functional managers, often based more on availability than on the actual needs of the team." To build high-performing teams, SD "managers need to ensure that users, as well as IS personnel, share a common understanding of the process to be followed, the scope of the project, and how the different parts of the team will link with each other." Leadership, the NPD research indicates, should be based on "the commitment to the team goal, irrespective of the function, [for that] can ensure less compartmentalization in software development activities between technologists and business (user) managers, and can lead to a more coherent product vision." The authors suggest that "functional managers are still held accountable for the resources assigned to teams from their departments," which puts their attention on controlling resources rather than performance. There is advantage to transforming the role of a functional manager "from a resource controller to that of a resource supplier" (pp. 213-214).

As global competition gives rise to increased demands to bring products quickly to market, NPD firms "have embarked on an umbrella of techniques and tools to accelerate product development. These include supplier involvement, overlapping phases, multifunctional involvement, and prototyping." An organization's ability to compete in terms of acceleration derives from organizational and process factors. "NDP literature emphasizes team composition as a primary factor in achieving development process acceleration." Specifically, the combination of "cross-functional teams and...a strong leader or champion" contributes greatly to acceleration. "Leadership style is another important predictor of success in process acceleration. The three most common leadership styles found in successful NPD environments are communicator, integrator, and planner." Management involvement—in terms of setting goals and reward systems, and effective communication—also plays a large role in NPD success. There are several process factors that also contribute to NPD success. These include "the use of tradeoffs and risk analysis, benchmarking, reuse, and the use of automated tools. Research in NPD shows that, of the five common tradeoffs used in development projects (cycle time, features, quality, cost of development, and product pricing), features and cycle time are perceived as the most important" (pp. 215-216).

In SD, development process maturity "refers to the degree of control exercised over the development process." NPD places its focus "more on the product rather than on the development process, and this lack of process discipline is often the most critical obstacle to successful NPD." SD has developed a "five-level capability maturity model (CMM)." It consists of "initial, repeatable, defined, managed, and optimizing" levels. Nambisan and Wilemon suggest that while "the key process areas defined in CMM may be unique to SD, the inherent structure of the maturity framework can be easily adapted to the NPD domain," and they are of value, even though CCM focuses "on the process definitions and parameters, and [does] not address the people and organizational issues that are equally important for organizations attempting to move from one maturity level to another" (pp. 216-217).

Implications of this research include its demonstrating "the promise for sharing successful practices and techniques in each of the two domains...stronger linkages between, [and that professional] organizations...[a need for] commitment in terms of resources for team formation and other team-building activities [and that] implementation of such cross-domain 'lessons' has several associated risks, and hence, the role of management in establishing a facilitating environment cannot be over-emphasized" (pp. 218-219).

Team Management and Software Development

Faraj and Sproull's 2000 *Management Science* article also concerns the role of teams in software development projects. For these authors, "software development teams are a significant example of the importance and the challenges of managing team-based knowledge work." The disparity between the amounts of resources invested in development and the large number of failed projects suggests that the nature of team development and management should be essential foci for any knowledge-management firm that is creating software: "Because software development is knowledge work, its most important resource is expertise." Given that development teams are most often formed de novo, and given the rarity of an experienced team moving, after completion, from one project to another, "savvy managers attempt to attract people with the most expertise to their projects" (p. 1554). Once the team is formed, however, savvy must give way to effective coordination and management, if the product is going to retain high quality and reach completion.

Expertise coordination is a subset of teamwork coordination, with expertise defined as "the specialized skills and knowledge that an individual brings to the team's task. Coordination refers to team-situated interactions aimed at managing resources and expertise coordination." Fajar and Sproull pose three research questions: "(1) What is expertise coordination? (2) What is the relationship between expertise coordination and team performance? (3) Does expertise coordination contribute to team performance above and beyond traditional factors such as group resources and the use of administrative coordination?" There are two basic views of expertise: "knowledge as abstract representation and expertise as the possession of that knowledge...[and] expertise as context-dependent, emerging from patterned interactions and practices in particular situations." The first view would posit that teams are ultimately "the aggregation of individual skills and knowledge," while the second suggests that a "team provides the context in which expertise can emerge through interaction" (p. 1555).

Fajar and Sproull propose that "expertise coordination consists of socially shared cognitive processes that develop and evolve in order to meet the demands of task-based skill and knowledge dependencies. When team members apply expertise to meet task demands, they activate and reinforce these processes." Moreover, "for expertise coordination to be effective, processes that are distributed, heedful, and emergent have to occur. These processes are distributed because expertise is dispersed among team members...heedful because overlapping task knowledge allows flexible and supportive joint action...and...emergent because answers and solutions are not prespecified but are generated through interactions" (p. 1556). To determine the validity of their three hypotheses—"conventional team factors are positively related to team performance...expertise coordination processes are positively related to team performance [and] expertise coordination processes are positively related to team performance above and beyond traditional factors" (p. 1558)—Fajar and Sproull collected data from 69 different teams that worked within application development departments of firms specializing in software development.

A combination of surveys and interviews provided the researchers with responses to weigh against findings in previous literature. For instance, prompts were developed from "three dimensions of expertise [that] are commonly recognized as important in software development: (1) technical expertise, (2) design expertise, and (3) domain expertise," allowing respondents "to evaluate, for each of these three dimensions of expertise, the percentage of necessary expertise that is located inside the team" (p. 1559). Data collected included years of experience, which is "often used as a proxy for domain knowledge and technical capability in software development," responses to questions about administrative coordination methods, software development methodologies used, and two "dimensions of [team] performance [that] appear essential for knowledge teams: effectiveness and efficiency" (pp. 1559-1560).

Faraj and Sproull's "findings suggest that previously inconclusive results regarding the link between coordination and performance may result from a lack of differentiation between efficiency and effectiveness dimensions of performance

[and] high levels of administrative coordination were unrelated to measures of team effectiveness. However, administrative coordination was strongly related to efficiency measures of performance, even after incorporating expertise coordination....The more fine-grained separation of performance into the two dimensions of efficiency and effectiveness allows a better understanding of which aspects of performance are affected by administrative coordination." There was only "weak support for the proposed relationship between the traditional factors and input variables on one hand, and team performance on the other....The presence of expertise did not affect any dimension of performance. Professional experience was found to have no impact on team effectiveness but was positively related to efficiency, indicating that more experienced teams are better able to control their budgets and schedules." Ultimately, "while expertise is a necessary input, its mere presence on the team is not sufficient to affect performance effectiveness if team members cannot coordinate their expertise" (p. 1564).

Knowledge Management and New Product Development

The combination of increased speed and accessibility of the Internet, and the increasing sophistication of software tools have created pathways for information and innovation that were unimaginable only two decades ago. Farris et al. (2003) report on "the Web of Innovation," which "permits an organization to benchmark and develop its own e-knowledge management program for the NPD [new product development] process." Their survey results indicate "that the most useful tools are not the most highly used, and commercial tools will usually require customization." The Web of Innovation model describes "the stages of knowledge transactions within the innovation process (e.g., creation, communication, usage, regeneration). It represents a formalization of the process of knowledge management within the R&D environment" (p. 24).

Knowledge flow is enabled by "culture, infrastructure and technology...Of these, culture is the most important as well as the most difficult to manage. Information Technology (IT) is perhaps the least important of the primary drivers but the easiest with which to develop commercial offerings to assist the NPD and other processes of the model." Even though these authors previously "delivered a strong caveat: IT tools are not in themselves knowledge management," NPD managers often look to IT first because of the "(supposed) ability of IT tools to enable the innovation process (i.e., improve the quality, speed to market and profitability obtained from the NPD process)" (p. 24).

The Web of Innovation model "consists of three levels, which reflect various activities, tasks and processes" that facilitate communication of information among the three levels. "The upper level reflects key business processes that drive, support or enable new product development. The second level reflects various types of enablers that support these R&D business processes through navigation, discovery, locating, analysis, interpretation, and searching....The foundation level reflects the ability to manage the explicit knowledge of the organization while proactively facilitating the sharing of tacit knowledge among individuals" (p. 26). Farris et al. make the point that the Web of Innovation model "mirrors the process of globalization," which they define as "having moved from penetrating markets around the world to learning from the world" (p. 25).

The first level of the Web of Innovation is concerned with process flow, and includes "the front end of innovation...those activities that come before the formal and well-structured new product and process development process (NPPD) with the identification and development of key opportunities for market, products and technology capabilities." Process flow also includes a technology strategy that identifies "core technologies and supporting skill sets needed to support business value and growth and integrate them into research programs and NPPD projects"; new product and process development: "a process through which products, processes or services translate capabilities into new value for the customers, organization or shareholders"; and portfolio management, "a decision process by which a collection of NPPD projects associated with a business are evaluated...and have their progress tracked to assure optimum time-to-market financial returns and growth in value of the business" (p. 28).

The second level consists of enabling processes and technologies, including portals, intellectual asset and competitive intelligence management, data and text mining and analysis, computer-aided analysis, the "systematic capture and archival of key organizational decisions and the reasons behind them," the processes used to "establish a balanced set of measurements...by which R&D processes are evaluated as contributing to the strategic direction and financial contribution to the organization...the ability to identify and utilize relevant expertise inside or outside of the organization," and collaboration in terms of sharing information, space, and tools. The third level, which Farris et al. term "foundation," consists of "knowledge capture, transfer and leverage" (pp. 28-29).

While using the model "for evaluating software tools used in the innovation process," Farris et al. report that besides finding "a plethora of tools being considered, developed, launched, and upgraded," there is also "a distinct lag between the aspirations of the software companies to build in functionality and the benefits being derived by industrial customers from those tools" (pp. 25-26). Quantification of benefits, both "hard" and "soft," "remains the Achilles heel of the knowledge flow process, including our Web-enabled tools." Hard benefits include "cycle time reduction and sales and earnings growth," and soft benefits include human factors. Conclusions reached include the point "that major corporations will continue to adopt IT tools from the outside or develop them internally in a piecemeal rather than an integrated fashion....The better IT tools will be those that support (and even tend to drive) the culture toward an improved innovation process....Culture remains the key to achieving knowledge flow, and thus to applying knowledge management techniques to generate profits" (p. 26).

Farris et al.'s survey results indicate that of the three categories of software tools used by respondents (those developed in-house, those that were commercially produced but customized, and those that were 'off-the-shelf'), "the tool category rated highest on usefulness (Data Mining with a 4.4) was among the least frequently used" while "some of the most-used tools were rated relatively low in their usefulness" (p. 31). Survey respondents also indicated that "successful implementation of a Web-enabled tool with their own organization...included observations on the amount of training needed to use the tool, its user-friendliness, and the value it provided for the investment, the quality of results, and its alignment with the organizational culture." These tools included those developed in-house and commercially, such as Lotus Notes, MS Project, MS Office, and Sci Finder. The tools cited "for least-successful implementation" were also produced commercially and in-house, and "interestingly," MS Project fell into this category as well. "This might be explained by the different ways that different organizations try to use that tool, by differences in organizational readiness for it, or by its different compatibility with different organizational cultures" (pp. 32-33).

OPEN SOURCE

Software development, as the previous sections discuss, involves management processes, innovative development practices, interdisciplinary theory and inter-departmental teamwork, and the advances that the computing sciences and technologies have made in natural language coding (object-oriented programming, particularly) and the Internet. Together, theory, commerce, and technology have spawned the ascension of knowledge as the currency of the 21st century. The proliferation of the open source movement may, therefore, be viewed as a curious—even paradoxical—strategy for organizations to employ, either fully or in part, as its fundamental tenet dismisses traditional copyright and patent actions for intellectual property developed cooperatively, iteratively, and recursively, with no monetary costs to the consumer, and challenges well-established new product development and organizational management strategies.

Open Source: An Economic View

Lerner and Tirole (2002) contend that the "interest in open source software development...has been spurred by three factors: the rapid diffusion of open source software...the significant capital investments in open source projects [and] the new organizational structure." They seek to answer what may be viewed as a conundrum: "Why should thousands of top-notch programmers contribute freely to the provision of a public good?" They explain their answers through the lens of economics, as they "seek to draw some initial conclusions about the key economic patterns that underlie the open source development of software." Indeed, the open source development process can be considered as "reminiscent of the time of 'user-driven innovation' seen in many other industries," and therefore look to research in new product development and other innovative venues as context for their economics-based discussion, as well as "the adoption of the scientific institutions...within for-profit organizations" (pp. 197-199).

Lerner and Tirole identify three periods or "distinct eras of cooperative software development," beginning with the early 1960s to the early 1980s, when "the sharing by programmers in different organizations of basic operating code of computer programs—the source code—was commonplace." The second "era," the early 1980s to the early 1990s, witnessed "the first efforts to formalize the ground rules behind the cooperative software development process." This included "a formal licensing procedure that aimed to preclude the assertion of patent rights concerning cooperatively developed software." In the third "era," which extends from the early 1990s to the current day, "widespread diffusion of Internet access...led to a dramatic acceleration of open source activity," permitting both "interactions between commercial companies and the open source community," and "the proliferation of alternative approaches to licensing cooperatively developed software" (pp. 200-202).

Lerner and Tirole ask three questions and use economic theory to answer them. "Why do people participate? Why are there open source projects in the first place? And how do commercial vendors react to the open source movement?" (p. 212). A programmer will participate if he or she "derives a net benefit (broadly defined) from engaging in the activity." Why would a programmer "forego the monetary compensation she would receive if she were working for a commercial firm or a university?" Benefits include the professional refinement of skills that come with the challenges that new software development projects present and the enjoyment that comes from working on a "cool" open source project, as opposed to "routine tasks" at work. There are also "delayed rewards" or incentives: "career concern incentives refers to future job offers, shares in commercial open source-based companies, or future access to the venture capital market. The ego gratification incentive stems from a desire for peer recognition" (pp. 212-213).

"Commercial projects have an edge on the current-compensation dimension because the proprietary nature of the code generates income." An open source project, however, "may well lower the cost for the programmer" in two ways. First, Lerner and Tirole refer to the "alumni effect; because the widely available free code can be used by educational institutions, programming code is often "already familiar to programmers." Secondly, there are "customization and bug-fixing benefits," which result from the collaboration of interested programmers on specific problems. Open source projects offer "delayed rewards (signaling incentive)." These include "better performance measurement," the result of contributors seeing all the code, not just what 'worked.' There is also "full initiative," meaning that the programmers involved maintain responsibility for the project, quite different from commercial projects in which employees are dependent on supervisory oversight. "Greater fluidity" is also a delayed reward: "Programmers are likely to have less idiosyncratic or firm-specific, human capital that limits shifting one's efforts to a new program or work environment" (pp. 212-216).

Quite often, individuals take on projects that "were motivated by information technology problems that they had encountered in their day-to-day work." Individual incentives also include "giving credit to authors." Overall, "the reputational benefits that accrue from successful contributions to open source projects appear to have real effects on the developers" (p. 218). In terms of project organization and governance, "Favorable characteristics for open source production are (a) its modularity...and (b) the existence of fun challenges to pursue." Leadership is essential: "The leader must provide a 'vision', attract other programmers, and last but not least, 'keep the project together" (p. 220).

Commercial software companies respond to the success of the open source movement by trying to "emulate some incentive features of open source processes in a distinctively closed source environment [and] to try to mix open and closed source processes to get the best of both worlds." This means increasing the visibility of key employees, resulting in a potential hiring away of top programmers, but also "enables them to attract talented individuals and provide a powerful incentive to existing employees." Internal sharing of code is another way that commercial companies try to emulate open source processes. "Strategies to capitalize on the open source movement [include] living symbiotically off an open source project [by] providing complementary services and products that are not supplied efficiently by the open source community." Also, a company can "take a more proactive role in the development of open source software" by releasing proprietary source code along with creating "some governance structure for the resulting open source process" (pp. 223-225).

Lerner and Tirole's examination of open source methodologies and their adoption in commercial environments concludes with suggested topics for further investigation, including the "role of applications and related programs": "As the open source movement comes to maturity, it will confront some of the same problems as commercial software does, namely the synchronization of upgrades and the efficient level of backward compatibility." The influence of a competitive environment suggests that "it would be useful to go back to the economics of cooperative joint ventures" when each of the members can come together against a dominant foe, their goals align (pp. 228-229). Open source, as an exportable strategy to other types of businesses, would be difficult to take root. Modularity, collaboration, and individual independence may be counter-productive in some industries where "the development of individual components require large team work and substantial capital costs, as opposed to (for some software programs) individual contributions and no capital investment" (p. 231).

Open Source: A Joint Management Model Perspective

von Hippel and von Krogh (2003) have a different perspective on the relationship between open source software development and the organization as compared to Lerner and Tirole's (2002) economic-based understanding. They suggest viewing open source software development as "an exemplar of a compound 'private-collective' model of innovation that contains elements of both the private investment and the collective action models and can offer society 'the best of both worlds' under many conditions." The private investment model "assumes that returns to the innovator result from private goods and efficient regimes of intellectual property protection. The 'collective action' model assumes that under the conditions of market failure, innovators collaborate in order to produce a public good" (von Hippel & von Krogh, 2003, p. 209).

Echoing the historical overview supplied by Lerner and Tirole (2002), von Hippel and von Krogh (2003) remind readers that "much of the software development in the 1960s and 1970s was carried out in academic and corporate laboratories by scientists and engineers. These individuals found it a normal part of their research culture to freely give and exchange software they had written, to modify and build upon each other's software both individually and collaboratively, and to freely give out their modifications in turn. This communal behavior became a central feature of 'hacker culture'." Such collegial generosity supporting the greater good gave way in the 1980s, "when MIT licensed some of the code created by its hacker employees to a commercial firm." In response, Richard Stallman, an MIT programmer who "viewed these practices as morally wrong impingements upon the rights of software users to freely learn and create...founded the Free Software Foundation in 1985, and set out to develop and diffuse a legal mechanism that could preserve free access for all to the software developed by software hackers. His pioneering idea was to use the existing mechanism of copyright law to this end." Software developers who wished to give open access of their source code to anyone interested in modifying it to suit their own purposes could take out a General Public License copyright. "Basic rights transferred to those possessing a copy of free software include the right to use it at no cost and the right to study its 'source code', to modify it, and to distribute modified or unmodified versions to others at no cost" (p. 210).

Given the increasing number of programmers, software development projects, and the growing consumer base for computer programs, "commercial software vendors [who] typically wish to sell the code they develop" found themselves in staunch opposition philosophically with some of the most talented programmers available. Moreover, "in current Internet days, rapid technological advances in computer hardware and software and networking technologies have made it much easier to create and sustain a communal development style at ever-larger scales" (p. 211). Given these developments and conditions, von Hippel and von Krogh suggest "that the practices of open source software developers and communities will be of interest to researchers working in organization science for two major reasons. First, open source software projects present a novel and successful alternative to conventional innovation models. This alternative presents interesting puzzles for and challenges to prevailing views regarding how innovation 'should' form and operate. Second, open source software development projects offer opportunities for an unprecedented clear look into their detailed inner workings" (p. 212).

By joining together the two dominant innovation models—private investment and collective action—developers create "a very rich and fertile middle ground where incentives for private investment and collective action can co-exist, and where a 'private-collective' innovation model can flourish." There are two ways in which private investment and open source models differ. "First, software users rather than software manufacturers are the typical innovators in open source. Second, open source innovators freely reveal the proprietary software they have developed at their private expense....However, manufacturers rather than product users have traditionally been considered the most logical private developers of innovative products and services because private financial incentives to innovate seem to be higher for them than for individual or corporate users....Individual user-innovators...can typically expect to benefit financially only from their own internal use of their innovations. Benefiting from diffusion of an innovation to the other users in a marketplace would require some form of intellectual property protection followed by licensing." Because there is no commercial market for products resulting from open source projects, "manufacturers' direct path to appropriating returns from private investment...often eliminates their incentive to innovate" (pp. 213-14).

The second difference between open source practice and private investment is the fact that users "in open source communities typically freely reveal their innovations by, for example, posting improvements and code on project Web sites where anyone can view and download them for free." This may seem incongruous, from the private investment point of view, as "users should only freely reveal their innovations when the costs of free revealing are less than the benefits." However, "a number of phenomena, ranging from network effects to increased sales of complementary goods, can actually increase innovators' private benefits if and as free revealing causes their innovations to be diffused more widely." Keep in mind that "there are two kinds of costs associated with revealing an innovation: those associated with the loss of proprietary rights to intellectual property, and the cost of diffusion" (p. 214).

Given that open source projects produce "knowledge and information products [in which] users [carry] out the entire innovation process for themselves—no manufacturer required...open source projects encompass the entire innovation process, from design to distribution to field support to product improvement. Such 'full function' user innovation and *production* communities are possible only when self-manufacture and/or distribution of innovative products directly by users can compete with commercial production and distribution. In the case of open source software this is possible because innovation can be 'produced' and distributed essentially for free on the Web" (p. 219).

The Maturing Open Source Movement

The journal *Research Policy* put out a special edition focused on the open source movement in 2003. Three of the articles, discussed in this concluding section of this chapter, pertain specifically to innovation phenomena resulting from the expansion of open source methodologies, participants, and product development. Lakhani and von Hippel (2003), for instance, comment that "explorations of the mechanics of and incentives to participate in open source software projects has focused on the core tasks of developing and debugging and improving the open source software itself." Lerner and Tirole's fundamental question remains: what drives top programmers to volunteer their time, knowledge, domain expertise, and goodwill to work on "these basic tasks"? Explanations include: "(1) a user's direct need for the software and software improvements worked upon; (2) enjoyment of the work itself; and (3) the enhanced reputation that may flow from making high-quality contributions to an open source project." These authors seek, therefore, to "understand why and how a task at the mundane end of the scale gets done," and why people choose to deliver "high-quality 'field support' to users of open source software" as the topic of their examination. (p. 923).

Lakhani and von Hippel "include altruism; incentives to support one's community; reputation-enhancement benefits received by information providers; and expectation of benefits from reciprocal helping behavior by others" among the motivating forces that drive programmers to answer "help desk" questions. To determine both the rationale and the profile of a field support volunteer, the authors "partition the overall task of information-providing into three subtasks: (1) the posting of a question by information seekers; (2) the reading of posted questions by potential information providers; and (3) the posting of answers," then embarked on empirical research on the Apache Web server software community. "We find that 98%, on average, of the time spent at the help Web site by an information provider is devoted to reading posted questions, and only 2% to providing answers. Information providers report that their motive for reading questions is primarily to learn about problems that other Apache users are experiencing....In other words, the major cost of providing help, matching of a posted question with a willing and able information provider, is carried out by providers because they directly receive a reward for this activity. The cost of actually answering questions, task (3), is generally very low, because providers only transfer information they already know to questioners, and report that they expend only 1-5 minutes on that task per answer provided." They do this for one of three reasons: "to promote open source software/free software movement...[because] they are motivated by an enhanced likelihood of receiving help...or by a sense of obligation from having received help from others in the past" (p. 924).

The authors' study of the Apache user group field support offers findings that "suggest that 'being a good company (open source project) citizen' and executing tasks 'important to the company (project)' may be important motives for participation. Enjoyment of the task of answering a question, 'part of my job' and reputational gains ('I enjoy earning respect') also appear, but less strongly" (p. 928). Information providers reported that "several types of benefit may be motivating [them] to respond." These include "both specific and generalized [expectations of] reciprocity"; strong identification "as belonging to a community...I will gain reputation or enhanced career prospects;...answering questions is intrinsically rewarding; [and] it is part of my job" (pp. 936-938).

Several implications can be drawn from this study, notwithstanding the lack of "robustness of the Apache help system...we have studied." The first is that the benefits of providing information "could be higher or lower in some open source projects, or that in a given project, this benefit may decrease if and as there is 'less to learn....Second, it is reasonable to ask whether the mechanism we have seen functioning well in the case of the Apache help Web site can also function effectively if question loads are much greater....Third, we found that low-cost provision of answers was possible in the case of Apache because some information providers could provide the requested information 'off the shelf'" (p. 939). Because the authors "found that the public posting of both questions and answers created a site that potential information providers wanted to visit and study in order to gain valuable information for themselves...it is important to analyze the micro-level functioning of successful open source projects to really understand how and why they work" (p. 940).

Open Source: Programmer Incentives and Organizational Benefits

Moving from Apache as subject and example to Freenet, "a project aimed at developing a decentralized and anonymous peer-to-peer electronic file sharing network," von Krogh et al. (2003) examine data to understand "the strategies and processes by which new people join the existing community of software developers, and how they initially contribute code." Their efforts result in "the constructs of 'joining scripts', 'specialization', 'contribution barriers', and 'feature gifts', and propose relationships among them" (p. 1217).

Citing Lerner and Tirole (2002), von Krogh et al. (2003) describe the "production of open source software (OSS) [as resulting] in the creation of a public good that is non-rival, i.e., users' utility from the software are independent, and non-exclusive, i.e., no individual or institution can be feasibly withheld from its usage." Acknowledging that "the success of a project in terms of producing the software relates to the growth in the size of the developer community," and that "joining a developer community may not be costless," these researchers are interested in creating a theory of "the open source software innovation process [that explains] how people sign up for the production of the public software good, under conditions where the cost of contributing vary." For instance, newcomers "share with existing developers greater benefits of revealing their innovations than those outside the community...because their ideas, bug reports, viewpoints and code can be reviewed and commented upon by other developers and users, and in terms of learning benefits, the group's feedback can be direct and specific to the newcomer" (pp. 1217-1218).

For the newcomer, the benefits of contribution, both immediate and future, far outweigh the costs. But coders' costs are not the same as those inherent in "commercial software development," which has a greater concern in concepts such as "modularization of software code," even though this yields similar benefits to open source software development such as an increase in the "project's transparency, lower barriers to contribute, and [allowance] for specialization by enabling efficient use of knowledge. Furthermore, efficiency in the innovation process requires that individuals specialize in certain areas of knowledge. Therefore, specialization in the developer community, by distinct software modules, could benefit the development process, and due to the barriers of understanding and contribution, this could be especially true for newcomers" (p. 1218).

Context for the Freenet project includes three important points. Because Freenet was developed from scratch, "joiners should have to put in considerable effort before they can contribute, and newcomers should not be able to realize immediate benefits from specializing in module development according to an official and pre-defined modular software architecture.…Secondly, Freenet was launched not on the basis of workable code…but on a master thesis in computer science written by its founder Ian Clarke outlining the theoretical principles of such a software system.…Thirdly, Freenet is a young project in comparison to the more established OSS projects like the Linux operating system that has been in operation since 1991 or the Apache Web server project which was founded in 1995." Researchers "gathered data from four different sources": structured telephone interviews, the project's archived mailing lists, "the history of changes to the software code available via the project's software repository within the *Concurrent Versioning Systems* (CVS) source code management tool," and fourth, "in order to obtain contextual understanding of the project we collected publicly available documents related to the project" (pp. 1219-1220).

Of the 356 participants in the developers e-mail list, 326 did not contribute code. Of the 30 who did, the researchers classified them as either: (1) joiner—someone who is on the e-mail list but does not have access to the CVS repository; (2) newcomer—someone who has just begun to make changes to the CVS repository; and (3) developer—someone who has moved beyond the newcomer stage and is contributing code to the project (p. 1220). The disparity between participants in the e-mail list and the actual number of contributors suggests the need for an "understanding [of] the mechanisms by which these developers join the project and the areas in which they contribute and specialize [as they are] crucial elements toward developing a theory of the open source software innovation process" (p. 1226).

Given that joiners not only have to demonstrate technical skill sufficient to participate but also an understanding of the expectations that the coding community holds for new participants, the researchers hold that "'joining' is a behavioral script that provides a structure for the activity of becoming a member of a collective action project.…We define a joining script as the level and type of activity a joiner goes through to become a member of the developer community. And therefore, joining scripts represents a cost to any would-be developer in the project.…Based on the data [and] analysis…we conclude with the following proposition: **Proposition 1.** *Participants behaving according to a joining script (level and type of activity) are more likely to be granted access to the developer community than those participants that do not follow the project's joining script"* (pp. 1227, 1229).

Positing that "there might be private and collective benefits resulting from specialization and division of labor in a project," the researchers "tracked the overall specialization in the developer community." Defining "'specialization' construct [as] measured by the address of code submissions, i.e., to which module of the software a particular contribution is made," the researchers differentiate between "'high' specialization [which] indicates that the same modules within the code base were changed over time by a developer, [and] 'generalization' [which] indicates multiple modules were changed by a developer"; they "found evidence of high specialization." In sum, while "high specialization allows for efficiency in innovation, there are also benefits to rotating people among jobs, in particular to broaden the understanding of a project, the increased sensitivity to coupling of tasks, and better management interfaces" (pp. 1229-1230).

Data collected led to the construct of 'contribution barriers,' described as consisting of: (1) "ease of modifying and coding module," (2) "the extent to which the potential developer can choose the computer language used to code for the module can vary," (3) "ease with which to 'plug' the module into the architecture," and (4) "the extent to which a module is intertwined or independently working from the main code." Data analysis provides the foundation for "**Proposition 2.** *In an evolving software architecture of open source software projects, contribution barriers of modules (modifying and coding, variation in computer language, plug-in, and independence) are related to the specialization of newcomers*" (pp. 1231, 1233).

'Feature gifts' is the third of four constructs resulting from the Freenet analysis. It "is measured by whether the first contribution is an extension or feature (sub module) in the reference model." Just like any other social exchange in which "individuals form relationships to maximize rewards and minimize costs...gifts are part of this process...open source software innovation hinges on contributors giving gifts in the form of code...in Freenet we found gift giving to increase early specialization of newcomers," leading to "**Proposition 3.** *Feature gifts by newcomers are related to their specialization in open source software projects*" (p. 1233). Given that the "early feature gift is...a way to quickly contribute code to the project [and] is consistent with studies of the innovation process where pre-existing domain knowledge or direct experience is a source of new product ideas and a source of user-to-user technical support," the authors offer "**Proposition 4.** *In open source software projects, feature gifts by newcomers emerge from the newcomers' prior domain knowledge and user experience.*" There are occasions when core programmers allow "other newcomers to make a fruitful contribution...although they may not have had intimate understanding of the evolving architecture," which leads to "**Proposition 5.** *Feature gift by newcomers are related to contribution barriers in an open source project*" (p. 1234).

There are several implications that this study has for future research on both "open source and commercial software innovation." First, whereas current "theorizing builds on the premise that all or most innovation will be supported by private investment and that private returns can be appropriated from this" obtains for commercial software, as well as that for open source, "the protection mechanisms partly or fully guarantee the rights of the user, by sustaining free revealing of software code...an extended theory [of innovation] must explain why, what, and how expert developers contribute for free to the production of a public good." Second, "more research should be devoted to test if the same patterns of newcomer specialization can be identified across a population of projects....Thirdly, future research should attempt to uncover how the evolution of the software architecture changes joining scripts....Fourthly, future research should attempt to investigate the production relationship between developer input and productivity....Fifth, [having] found evidence in commercial Internet software development that higher performance of the development project was associated with the use of development teams with greater knowledge from several releases of a particular software...research should investigate if projects differ much in terms of turnover among various contributors, as well as what factors impact on turnovers" (pp. 1235-1236).

Bridging Commercial and Open Source Methodologies

West (2003) addresses the processes and impacts of open source software development and innovation approaches that were instituted in three of the largest proprietary names in computing—Apple, IBM, and Sun Microsystems—to determine the effectiveness of hybridization between open source and commercial methodologies in competitive proprietary environments. Given the introduction of Linux, the open source operating system, as a challenger to both Unix and Windows, there is now not only competition "to control a platform's direction," but a clear exemplar of how the various development "strategies reflect the essential tension of de facto standards creation: that between appropriability and adoption. To recoup the costs of developing a platform, its sponsor must be able to appropriate for itself some portion of economic benefits of that platform. But to obtain any returns at all, the sponsor must get the platform adopted, which requires sharing the economic returns with buyers and other members of the value train. The proprietary and open source strategies correspond to the two extremes of this trade-off [which means] leading computer vendors face a dilemma of how much is open enough to attract enough buyers while retaining adequate returns" (p. 1259).

A proprietary platform "consists of an architecture of related standards, controlled by one or more sponsoring firms." West refers to Teece (1986) and the discussion of appropriability, which "suggests that firms must use some combination of speed, timing and luck if they hope to appropriate returns generated by their innovation, [which] require the provision of complementary assets to commercialize. When additional investment is required to co-specialize the asset to be useful with a given innovation, the successful adoption of the innovation and related assets are mutually reinforcing, providing a positive feedback cycle" (West, 2003, p. 1260). However, creating and maintaining a proprietary platform does not always result in profiting from technology innovation. "When competing firms control different layers of the standards architecture, platform leadership is unstable because control of the platform can shift without disrupting the buyer's value proposition" (p. 1261).

West provides a comprehensive overview of the history of the three largest and most successful operating systems firms' strategies for building mainframes and complementary assets such as software applications, and points out that opting for one platform meant incurring significant costs upon conversion to another, as each platform had software that was "specialized to fit each platform's APIs" (p. 1261). With the rise of the personal computer industry, "the cost of entry...led to platform convergence as a large number of systems makers purchased processors from a shrinking number of microprocessor vendors.....IBM's PC architecture used both a processor and operating system from outside vendors. When coupled with its unexpected legal defeats on ROM copyrights, IBM lost control of its platform as other firms produced 'clone' computers that ran the same application software." IBM never regained dominance of the market, even after "spending billions of dollars on proprietary technologies in an unsuccessful bid to reassert its leadership" (p. 1262).

The Unix workstation, put out by AT&T in 1969, "eventually evolved into a new form of non-proprietary platform standard, often referred to as the 'open systems' movement." It shared with Microsoft's MS-DOS, however, its strategy of "licensing their operating systems to multiple hardware vendors, each [making] their platform ubiquitous by reducing switching costs and differentiation between hardware vendors. As with even the most proprietary computing platform, UNIX and MS-DOS were 'open' to third-party software suppliers, utilizing APIs widely disseminated to maximize software availability" (pp. 1262-1263).

Open source, according to West, got its start in the 1980s through Unix, which "had three main attractions for programmers: it ran on inexpensive minicomputers, was hardware independent and provided a state of the art environment for software development," including the most prominent example, Linus Torvald's Linux operating system. Richard Stallman's Project GNU extended "by analogy the traditions of pooled scientific research and the free dissemination of ideas, [arguing] that all software should be 'free software', with source code that can be read, modified and redistributed," and UC Berkeley's BSD Unix, which "solicited outsiders to volunteer to rewrite components...using only published APIs," eventually resulted in "firms that sold Linux- and GNU-related support and services...[labeling] their common vision 'open source'" (p. 1265). West's study exposes Apple, IBM, and Sun's turn to open source "to revitalize their strategies" after suffering downturns in their "historically...vertically integrated platforms differentiated by proprietary software....All three faced serious competitive pressures from Microsoft, and all three faced a challenge in formulating 'open' strategies that nonetheless allowed them to retain one or more sources of competitive advantage" (p. 1266).

Apple tried but failed to develop a competitive PC operating system and turned to Linux, "but it was eventually replaced by a new OS that combined a unique mix of proprietary and open source components." With the release of Mac OS X, "Apple proclaimed it was 'leading the industry by becoming the first major OS provider to make it's [sic] core operating system available to open source developers," yet Apple held "some layers of its operating system entirely proprietary....By opening only part of its technology...Apple made it less valuable to user-contributors. The fewer users that contributed...the less benefit Apple realized from its open source strategy" (pp. 71-72).

"IBM's development of applications using open source software had three common threads. First, IBM accepted commoditization of certain layers of its application architecture and...thus was willing to collaborate with open source software programmers to make a shared technology available to all...Second, in many cases the shared software competed with proprietary solutions developed by Microsoft...Finally, the shared software was released under non-GPL license allowing IBM to retain technology or make proprietary enhancements" (p. 1273). "Sun successfully differentiated itself from other rivals through ongoing enhancements in its operating system, particularly with its support for data networks....To improve adoption, it licensed its workstations and OS technology to customers and competitors [while it] retained full control of the architecture, allowing it to rapidly evolve the technology rather than negotiate with standards committees. As such, Sun's strategy more closely fit the 'proprietary but open' model of Morris and Ferguson (1993) than did Microsoft" (West, 2003, p. 1274).

"The experience of Apple, IBM and Sun suggests that shifting to even a partly-open architecture may require a major external shock to force firms to relinquish previous innovation-driven differentiation strategies [such as] opening parts—waving control of commodity layer(s) of the platform, while retaining full control of other layers that presumably provide greater opportunities for differentiation; [and] partly open—disclosing technology under such restrictions that it provides value to customers while making it difficult for it to be directly employed by competitors" (p. 1279). "Waving intellectual property rights makes the [platform-related] standard (or implementation) more attractive to competitors and key users, priming the positive-feedback bandwagon effects that can accrue to early market leaders. It also increases the number of products that are interoperable with the vendor's products" (p. 1280).

Given that open source "is best suited for technically proficient users...with strong motivations for customization...The degree to which open source adds value beyond this niche depends on how much it enables other attributes more directly valued by users, such as greater reliability, lower cost or expanded variety of complementary assets." West suggests that future research ask the following questions: "Is forfeiting IPR [intellectual property rights] more or less desirable in cases where competitors (or substitutes) have weak appropriability? Is such a differentiation strategy more effective when competing with an organization with tight IPR control? Or does it have general applicability as a strategy to pre-empt adoption of competing technologies? Is…the effectiveness dependent on the nature of the technology, its use, or the importance of complementary assets? Are there any other adoption process characteristics relevant to deciding on such strategies?" (p. 1282).

Suggested questions for future research particularly relevant to technology innovation management include: "Under what conditions will open standards lead to the widespread adoption of open source implementations by commercial vendors? How durable are various approaches (including opening parts, partly open) for creating barriers to imitation using open source software? Is the decision between open source and proprietary strategies a matter of managerial discretion? Or is there a normative 'best practice' for similar situated firms that dictate the appropriate choice for each context?" (p. 1283).

CONCLUSION

The use of software for the development of innovative product development and business practices has been an ongoing enterprise since automated data processing became a standard element of commercial enterprise half a century ago. The demand for productivity software feeds an ever-increasing proliferation of modifiable, tailored, and powerful software applications, which in turn created the need not only for machines and applications to extract and analyze data, but for management practices that foster the ease, growth, and profitability of people and systems using hardware and software, as well as the explosion of the a software engineering industry.

Boehm and Ross (1989) offer a project management theory, Theory W, to help software development industry managers identify all the steps to take to produce an outcome in which everyone involved considers him- or herself a winner. Boehm (1991) suggests that software development project managers employ decision and risk management trees as risk management techniques for software development projects, as managers need to focus on critical success factors and thereby mitigate risk. Fichman and Kemerer (1997) are also concerned with software process innovations, finding that "organizations will have a greater propensity to innovate with complex technologies when the burden of organizational learning is effectively lower, either because the organizations already possess much of the know-how and technical knowledge necessary to innovate, or because they can acquire such knowledge more easily or more economically" (Fichman & Kemerer, 1997, pp. 1345-1346). Nambisan and Wilemon (2000) describe the potential benefits derived from interdisciplinary sharing of knowledge and theories between the fields of technology innovation and software development. They find value in researchers viewing "product development along three dimensions—*people, process, and technology*" (p. 211).

Faraj and Sproull's (2000) focus is on the role of teams in software development projects, finding that once a team is formed, effective coordination and management are essential if the product is going to retain high quality and reach completion. Farris et al. (2003) conclude "that the most useful tools are not the most highly used, and commercial tools will usually require customization." Their Web of Innovation management model describes "the stages of knowledge transactions within the innovation process (e.g., creation, communication, usage, regeneration) [and] represents a formalization of the process of knowledge management within the R&D environment" (p. 24).

Open source software development—its history, proliferation, and current strong activity, also involves management processes, innovative development practices, interdisciplinary theory, and interdepartmental teamwork. Using an economic lens on this development phenomenon, Lerner and Tirole (2002) identify value in the motivating factors that drive individuals to participate in open source projects, calling for further investigation into how modularity, collaboration, and individual independence might prove beneficial for commercial software producers.

von Hippel and von Krogh (2003) suggest viewing open source software development as "an exemplar of a compound 'private-collective' model of innovation that contains elements of both the private investment and the collective action models and can offer society 'the best of both worlds' under many conditions." The private investment model "assumes that returns to the innovator result from private goods and efficient regimes of intellectual property protection. The 'collective action' model assumes that under the conditions of market failure, innovators collaborate in order to produce a public good" (p. 209). By joining together the two dominant innovation models—private investment and collective action—developers create "a very rich and fertile middle ground where incentives for private investment and collective action can co-exist, and where a 'private-collective' innovation model can flourish" (p. 213). Lakhani and von Hippel (2003) suggest that rather than focusing on core tasks of software development, researchers should seek to "understand why and how a task at the mundane end of the scale gets done," and why people choose to deliver "high quality 'field support' to users of open source software" (p. 923). By doing so, researchers will find that the benefits of providing information are variable across projects and, that programmers participate for their personal edification and potential benefit.

von Krogh et al. (2003), in their examination of the reasons and methods that software developers employ when involving themselves in open source projects, obtain similar findings to Lakhani and von Hippel (2003), though their focus was on user support functions. West (2003) addresses the processes and impacts of open source software development and innovation approaches that were instituted in three of the largest proprietary names in computing—Apple, IBM, and Sun Microsystems—to determine the effectiveness of hybridization between open source and commercial methodologies in competitive proprietary environments. Their history of merging proprietary and open source methodologies leads West to suggest areas for future research particularly relevant to technology innovation management, including the conditions that will support adoption of open source implementations by commercial entities, the durability of techniques for barring against imitation, the factors that determine development strategies, and the nature of best practices given the scope of development contexts.

REFERENCES

Attewell, P. (1992). Technology diffusion and organizational learning: The case of business computing. *Organization Science, 36*(1), 1-19.

Boehm, B.W. (1991). Software risk management: Principles and practices. *IEEE Software, 8*(1), 32-41.

Boehm, B.W., & Ross, R. (1989). Theory-W software project management: Principles and examples. *IEEE Transactions of Software Engineering, 15*(7), 902-916.

Downs, G., & Mohr, L. (1976). Conceptual issues in the study of innovation. *Administrative Science Quarterly, 21*(December), 700-714.

Faraj, S., & Sproull, L. (2000). Coordinating expertise in software development teams. *Management Science, 46*(12), 1554-1568.

Faris, G.F., et al. (2003). Web-enabled innovation in new product development. *Research Technology Management, 46*(6), 24-35.

Fichman, R.G., & Kemerer, C.F. (1997). The assimilation of software process innovations: An organizational learning perspective. *Management Science, 43*(10), 1345-1363.

Fisher, R., & Ury, W. (1983). *Getting to yes.* Baltimore, MD: Penguin.

Lakhani, K.R., & von Hippel, E. (2003). How open source software works: "Free" user-to-user assistance. *Research Policy, 32,* 923-943.

Lerner, J., & Tirole, J. (2002). Some simple economics of open source. *Journal of Industrial Economics, L*(2), 197-234.

Meyer, A., & Goes, J. (1988). Organizational assimilation of innovations: A multi-level contextual analysis. *Academy of Management Journal, 31*(4), 897-923.

Morris, C., & Ferguson, C. (1993). How architecture wins technology wars. *Harvard Business Review, 71*(2), 86-96.

Nambisan, S., & Wilemon, D. (2000). Software development and new product development: Potentials for cross-domain knowledge sharing. *IEEE Transactions on Engineering Management, 40*(2), 211-220.

Raymond, E. (1999). *The cathedral and the bazaar: Musings on Linux and Open Source by an accidental revolutionary.* Cambridge, MA: O'Reilly.

Teece, D. (1986). Profiting from technological innovation: Implications for integration, collaboration, licensing and public policy. *Research Policy, 15*(6), 285-305.

von Hippel, E., & von Krogh, G. (2003). Open source software and the "private-collective" innovation model: Issues for organization science. *Organization Science, 14*(2), 209-223.

von Krogh, G., Spaeth, S., & Lakhani, K.R. (2003). Community, joining, and specialization in open source software innovation: A case study. *Research Policy, 32,* 1217-1241.

West, J. (2003). How open is open enough: Melding proprietary and open source platform strategies. *Research Policy, 32,* 1259-1285.

Chapter XI
Directions in the Field of Technology Innovation Management

INTRODUCTION

Although the goal of this book is to provide foundational knowledge through in-depth consideration of the seminal literature in the technology innovation management field, we now offer some thoughts on integrating the past, present, and future research directions in this field. The underlying theme that holds together the research considered in this book is the tension between the old (current routine) and the new (innovation). Mainstream business and management theory, like economic theory, focuses on the assumption of equilibrium. The study of technology innovation management at its core considers how to manage in the face of dynamics caused by the novelty and uncertainty associated with innovation. The nature of these dynamics can differ depending on a variety of factors. In some cases, the innovation causes smaller disruptions, due either to the magnitude or the nature of its effects. Such changes are often associated with terminology such as continuous, evolutionary, incremental, or sustaining. At other times, the disruptions are quite large, either due to a greater magnitude of change or a substantial difference in the change. These changes are often associated with terms such as discontinuous, disruptive, radical, or revolutionary. A major challenge to technology innovation management research is that the assumption of equilibrium is needed in many cases to allow for sufficient simplification of phenomena to produce generalizable theory and solutions that are tractable and close formed.

While it is difficult to produce elegant theory and formulations in a dynamic environment, it is still possible. Even if it were not possible, research in this area is still worthwhile to conduct. While it may be possible for academic researchers to overlook inconvenient phenomena for the sake of simplifying reality to develop theory, practitioners must make decisions in both the presence and absence of theory. Consequently, no matter how context dependent or limited in explanatory power the early attempts at theory are, these attempts are still worthwhile, since they are a step forward in assisting practitioners in moving beyond intuition. For many researchers the need to support practitioners (managers) in a highly applied field (business and management) is a suitable call for research efforts. For those whose outlook is specifically on basic science, the feeling might be, *Why should we place our efforts into helping practitioners? This is the job of consultants—not research academics.* In this case, it is worth considering that research on technology and innovation management is still extremely worthwhile, since while theory exists, competing concepts and perspectives jostle for recognition and acceptance. This current status describes a field that is at the focal point of Kuhn's view of scientific revolutions (1962). That is, disagreement on theory exists because the field of technology innovation management currently sits in the messiest part of scientific evolution—a scientific revolution. Unsurprisingly, it is not clear when our field will emerge from this state. However, it places this sub-field of business and management research apart from many other areas of study, where the focus is currently on traditional science—incremental change driven by empirical studies.

Having just considered the status of technology innovation management research from an evolutionary standpoint and having considered the seminal works in this field throughout the text, we will turn our attention to more recent events. While it is not possible to state with any certainty which recent research will be considered seminal work several decades from now, it is possible to give insights into current trends in research and to project these out into the future. The consideration of open source was an attempt to make a prediction in the massive sub-field of information systems. Time will tell whether that subject and the articles discussed in this section become seminal.

Where Is Seminal Work Published and Why?

This leads to the question: where does seminal work come from? Based on our observations of the most frequently cited works in Technology and Innovation Management, most of the articles came from mainstream management journals as opposed to specialty journals. Clearly, as will be discussed further below, many articles that appear in a specialty journal in any field have an appeal that is too narrow to fit in a broader scope management journal. As a result, an innovation management paper that makes a contribution such that it is able to fit in a journal with a broader mission and coverage is more likely to be incorporated in articles in the broader field as well as the specialty journals. For example, articles that have been considered in this book that were published in the *Academy of Management Journal* are cited later on in both journals with a broad mission—like the *Academy of Management Journal*—as well as specialty journals.[1] In the past, this may have been due in part to the differential reach between specialty and broader mission journals. Before search engines and electronic access to journals, many researchers would go to libraries and look at the table of contents of journals they considered relevant. Consequently, a marketing researcher may have looked at journals such as the *Journal of Marketing* and *Journal of Marketing Research* every month. Although an innovation management researcher who approaches innovation research from a marketing perspective would look at the same group of journals as the marketing researcher, he or she might also look at some specialty journals, such as the *Journal of Product Innovation Management.* The result in the past was a much larger audience for the discipline-specific broad focus journals and a smaller market for the technology innovation management specialty journals.

However, with the proliferation of online databases, researchers are more likely to search keywords now rather than look at journal tables of contents or flip through the pages of their favorite journals. The search engines to a great extent enhance researchers' likelihood of finding an article. If a journal is listed on the major search engines, then people will find articles that are relevant to them when they conduct a literature search—regardless of whether the article appears in a specialty journal or a journal with a much broader focus. If a journal is listed on the search engines that also record citation counts, then one will be able to determine the relative level of use of different articles. Whereas in the past it could be very important to place an article in one of a very few *correct* journals, now the number of *correct* journals is more a function of the reach of the search engines.[2] Having said this, there is still a perception that an article is likely to be more important if it appears in certain journals.[3]

Another reason for an article in a mainstream journal being cited more heavily than an article in a specialty journal could be that the innovation-relevant article in a mainstream journal is written in a way to appeal to both the mainstream reader and the specialist. Consequently, there are two audiences that are likely to cite the article. The specialty journal article, however, may have less of a mainstream appeal—either in terms of topic or in the manner it is written. There are other reasons, such as perhaps the articles in specialty journals are (1) too focused, (2) not as information-rich as articles in journals with broader missions, or (3) perceptions that journals with broader missions discourage citing specialty journals.[4] It is also worth noting that books are often related to one or more seminal papers. In some cases, the paper is written prior to the book, while in other cases, the paper is follow-up work to the book. We see this, for example, in some of the books mentioned in Chapter I, in that they were followed up in later chapters with a discussion of a seminal journal article on the same or similar subject.[5]

Having briefly considered some of the non-content-related factors that are relevant to literature becoming seminal in the past, the next step is to try to understand what is of future importance. To do this, the topics that have been considered recently in the technology innovation management specialty journals are considered. To minimize the pitfalls of personal biases, this examination will be conducted in a very structured way.

Method

The abstracts of the 200 most recent articles were collected from the 10 technology innovation management specialty journals listed on the science and/or social science citation index (see Table 1). A count of individual and pairs of words was conducted to identify the frequency of occurrence. This list of words was processed to remove non-subject-specific words, such as *a, the,* and *and.* By putting together such a list, we can see what is important in current technology and innovation management research.

To gain deeper insight into how technology innovation management research differs from research in other fields of business and management, the list is compared to a similar list of words from all management journals listed on the science and/or social science citation index. The technology and innovation management specialty

Table 1. Technology innovation management specialty journals considered in this study

Journal Name	Index
IEEE Transactions on Engineering Management	SCI/SSCI
International Journal of Technology Management	SSCI
Journal of Engineering and Technology Management	SSCI
Journal of Product Innovation Management	SSCI
R&D Management	SSCI
Research Policy	SSCI
Research Technology Management	SSCI
Technological Forecasting and Social Change	SSCI
Technology Assessment and Strategic Management	SSCI
Technovation	SCI, SSCI

(SCI—Science Citation Index; SSCI—Social Science Citation Index)

journals are a subset of this list of about 200 journals. Business and management journals consist of all journals that are listed in the subject areas of: business, management, economics/finance, or operations research and management science. To this list, the journals on the Financial Times 40 List of Management Journals were added.[6] Journals that did not have abstracts available on either the Web of Science or Scopus were removed from the list.[7]

Results of Study

Word frequency between all management journals and technology innovation management specialty journals clearly differs (see Appendix I). Due to the lack of context that a single word offers, it is difficult to get much insight from these sorts of differences. Many of the differences are unsurprising—words related to technology and innovation appear high on the list of most frequently used words in the abstracts from the specialty journals and much lower down on the list if one considers all business and management journals.

More insights are offered on the list of the top 100 word pairs. Many of the word pairs found frequently in the technology innovation management journals emphasize some relationship to the concepts of technology or innovation, but there are other notable differences from the entire group of management and business journals. The general field of business and management places relatively more attention on *information technology* and *information systems,* but gives less attention to *knowledge management* and related concepts such as *knowledge based* and *knowledge transfer.* Many different financial terms appear frequently in business and management journals, while in the technological innovation management specialty journals, the financial focus is limited to discussions of *venture capital* and *real options.* In technology and innovation management specialty journals, marketing-related discussion focuses on the development of new products—terms such as *product development, new product(s),* and *development process* are used. In the field of human resources, the technology innovation management journals tend to lean more heavily on the concept of *human capital,* as opposed to *human resource.*

The technology innovation management journals also differ from other journals in that their focus on concepts related to the interaction of multiple organizations is increasing, while the term *supply chain management* is frequently used by both groups of journals studied—more frequently in fact by all journals. Other terms associated to inter-relation between organizations such as *strategic alliance, joint venture, regional innovation,* and *science parks* appear more frequently in the technology innovation management specialty journals.

Other notable differences are the focus on the *long term* instead of the *short term* in technology innovation management journals. Areas of recent focus in technology innovation management specialty journals include *cross functionality, technology transfer,* and *intellectual property.* There is a strong focus on various aspects of policy in the technology innovation management journals. What is noteworthy is that *developing countries* also appear to be given much more attention in technology innovation specialty journals than in other types of business and management journals. This difference is quite surprising, because developing countries are historically viewed as being laggards in research, technology development, and innovation. While there are many similarities in the types of single words and word pairs that arise in the recent overall business and management literature and the technology innovation management literature, there are also clear differences. The similarities raise the possibility for the identification of topics that will span both areas, while the differences highlight areas where the technology innovation management specialty journals are likely to make contributions to the field of management that are unique.

Future Directions for Consideration

Areas such as new product development have been the focus of much research activity for years now. One could possible argue the same for technology transfer. However, the importance of technology transfer is likely to grow both in volume and scope. The increasing focus on supply chains and the increase in research in network or relationship-related concepts in the technology innovation management specialty journals suggests that new emphases and approaches to technology transfer research might be close at hand. This increased focus on supply chains and networks will also allow for new perspectives on existing questions and the development and testing of new models.

The past emphasis on large firms in developed economies, while still important, is declining as greater focus is placed on technology and innovation management in emerging and developing economies. In many cases consideration, here, focuses on small firms. It is likely that there will be a reemergence in discussion of appropriate technology (as opposed to high tech) in response to differing needs in transition and emerging economies.

While technology innovation management research has traditionally focused on manufacturing, the growing realization that innovation in services can create tremendous value is resulting in increased interest. Due to the relatively small number of researchers with established research programs in service innovation,[8] it will be at least a decade before this field matures.

Increasing concern regarding the natural environment and energy is also likely to greatly affect technology and innovation management work in the future. Discussions of global warming and adoption of sustainability as a policy by local and national governments will lead to further consideration of this area.

There is a lack of agreement in the innovation literature on whether activities are occurring at an increasing rate. The increased rate of knowledge generation due to the involvement of more actors and the automation of certain types of data collection and analysis will ensure that the question of time and rate will remain critical. In addition, the issue of the management of convergence of different technologies will become increasingly topical. Of course, discussion relating to the future is rich in speculation, and as any researcher in technology and innovation management will tell you, quite risky.[9]

REFERENCES

Kuhn, T.S. (1962). *The structure of scientific revolutions.* Chicago: University of Chicago Press.

Schnaars, S.P. (1989). *Megamistakes: Forecasting and the myth of rapid technological change.* New York: The Free Press.

ENDNOTES

1. Note: highly cited articles in specialty journals, such as *Research Policy,* are rarely or not at all cited in journals with a broad mission.
2. If one wants to be able to demonstrate the impact of articles, then placing articles in journals that are tracked by search engines that also conduct citation counts is beneficial.
3. The mode number of cites for articles regardless of the reputation of a journal is typically 0. In terms of impact and contribution, the quality of the article as opposed to the quality of the journal is what is critical.
4. While I have heard a number of authors express this opinion in different ways, I have not heard editors express this in relation to their journals. So this point is not offered as a fact, but in an effort to show that there are many possible different reasons that may or may not explain past and future occurrences.
5. Authors like Allen, Christensen, Prahalad and Hamel, and von Hippel are examples of this occurrence.

[6] Except for a few journals such as the *Journal of American Statistical Association, Management International Review,* and *Econometrica,* the journals are already on this list.

[7] For example, *Fortune* appears on the social science citation index, but no abstracts are supplied.

[8] There are of course exceptions; the primary exception is innovation research in management information systems—a field with a well-developed theoretical base and research community.

[9] See, for example, Schnaars (1989).

APPENDIX A

Most frequently cited words from management and business journals vs. technology innovation management specialty journals

All Journals	Technology Innovation Management Journals
model	technology
results	management
performance	type
time	partner
firms	innovation
new	document
data	inward
market	new
management	development
information	product
problem	united
process	firms
system	cited
rights	paper
analysis	business
business	knowledge
reserved	process
approach	study
different	based
firm	these
value	industry
product	times
show	technological
models	between
systems	performance
design	time
organizational	model
development	information
risk	project
knowledge	market
problems	firm
level	analysis

continued on following page

APPENDIX A CONTINUED

method
effects
work
price
cost
service
theory
find
quality
case
control
important
decision
first
technology
number
optimal
proposed
implications
relationship
effect
factors
studies
impact
literature
framework
managers
industry
financial
algorithm
provide
empirical
related
social
support
change
evidence

systems
rights
reserved
results
projects
companies
case
technologies
strategic
policy
system
industrial
design
different
data
use
strategy
high
review
organizational
factors
learning
using
products
national
change
processes
important
through
manufacturing
value
role
technical
institute
engineering
level
success

continued on following page

APPENDIX A CONTINUED

role	managers
policy	activities
strategy	empirical
significant	approach
network	studies
methods	economic
costs	framework
customer	growth
developed	three
behavior	impact
present	most
function	innovations
rate	however
economic	software
processes	strategies
companies	decision
future	implications
those	support
organizations	social
structure	company
strategic	marketing
production	competitive
markets	organization
sample	future
learning	network
manufacturing	quality
general	first
strategies	relationship
variables	industries
supply	related
conditions	literature

APPENDIX B

Most frequently cited words for management and business journals vs. technology innovation management specialty journals:

All Journals	Technology Innovation Management Journals
supply chain	product development
decision making	new product
business media	case study
product development	technology management
information technology	relationship between
time series	technology transfer
exchange rate	information technology
corporate governance	high tech
system dynamics	supply chain
information systems	knowledge management
competitive advantage	decision making
optimization problems	project management
short term	technology based
customer satisfaction	new technology
service quality	development process
interest rate	new products
optimization problem	high technology
human resource	information systems
linear programming	case studies
knowledge management	competitive advantage
objective function	long term
interest rates	new technologies
stock market	product innovation
cross sectional	development projects
Monte Carlo	cross functional
job satisfaction	technological change
real time	technological innovation
organizational behavior	intellectual property
scheduling problem	innovation management
organizational learning	innovation systems
rcal world	industrial research
monetary policy	innovation process

continued on following page

APPENDIX B CONTINUED

top management
public relations
proposed model
developing countries
decision support
quality management
human capital
optimal control
financial performance
life cycle
supply chains
genetic algorithm
small firms
chain management
behavior management
new products
NP hard
stock price
non linear
stock returns
transaction costs
integer programming
firm level
mixed integer
operational research
discrete time
organizational change
market share
neural network
resource based
market orientation
dynamic programming
polynomial time
steady state
optimal solutions
manufacturing firms
social capital

technology policy
research institute
developing countries
human capital
development NPD
strategic management
knowledge transfer
knowledge based
firm performance
product design
resource based
empirical study
innovation system
market orientation
manufacturing firms
technology development
empirical evidence
intellectual capital
organizational learning
technology strategy
social capital
strategic planning
empirical analysis
public policy
success factors
medium sized
innovation document
industrial engineering
relationships between
science parks
human resource
radical innovation
life cycle
absorptive capacity
pharmaceutical industry
chain management
disruptive technologies

continued on following page

APPENDIX B CONTINUED

organizational performance
conceptual framework
cash flow
second order
development process
risk management
exchange rates
sensitivity analysis
transaction cost
sufficient conditions
firm size
financial intermediation
international business
lower bound
management practices
resource management
Web site
strategic management
health care
venture capital
Web based
model based
business ethics
economic growth
continuous time
public policy
high tech
closed form
high quality
electronic commerce
dynamics review
science park
technology intelligence
co operation
research center
small firms
team members
regional innovation
product success
joint ventures
innovation policy
software development
strategic alliances
technological innovations
technological development
semiconductor industry
research institutes
economic growth
technological capability
product quality
knowledge intensive
development performance
technological learning
management practices
study jet
project performance
quality management
firm level
core competencies
venture capital
real options
cycle time

APPENDIX C

Table 1. Original articles sorted by citation counts

Lead Author	Yr.	Journal	Times Cited 2004
COHEN	1990	ADMINISTRATIVE SCIENCE QUARTERLY	815
PRAHALAD	1990	HARVARD BUSINESS REVIEW	590
TUSHMAN	1986	ADMINISTRATIVE SCIENCE QUARTERLY	541
TEECE	1986	RESEARCH POLICY	487
HENDERSON	1990	ADMINISTRATIVE SCIENCE QUARTERLY	425
DOSI	1982	RESEARCH POLICY	420
DAVIS	1989	MANAGEMENT SCIENCE	341
LEONARD-BARTON	1992	STRATEGIC MANAGEMENT JOURNAL	314
PORTER	1985	HARVARD BUSINESS REVIEW	306
IVES	1984	MANAGEMENT SCIENCE	281
NELSON	1977	RESEARCH POLICY	267
PAVITT	1984	RESEARCH POLICY	266
BARLEY	1986	ADMINISTRATIVE SCIENCE QUARTERLY	265
KIMBERLY	1981	ACADEMY OF MANAGEMENT JOURNAL	250
DAMANPOUR	1991	ACADEMY OF MANAGEMENT JOURNAL	249
POWELL	1996	ADMINISTRATIVE SCIENCE QUARTERLY	236
GRANT	1996	STRATEGIC MANAGEMENT JOURNAL	234
MARCH	1991	MANAGEMENT SCIENCE	226
BANTEL	1989	STRATEGIC MANAGEMENT JOURNAL	222
DOWNS	1976	ADMINISTRATIVE SCIENCE QUARTERLY	217
ZMUD	1979	MANAGEMENT SCIENCE	214
MAHAJAN	1990	JOURNAL OF MARKETING	213
COOPER	1979	JOURNAL OF MARKETING	210
ROTHWELL	1974	RESEARCH POLICY	204
FISHER	1971	TECH FORECAST & SOCIAL CHANGE	199
HAGE	1969	ADMINISTRATIVE SCIENCE QUARTERLY	198
ANDERSON	1990	ADMINISTRATIVE SCIENCE QUARTERLY	190
ABERNATHY	1985	RESEARCH POLICY	188
DOUGHERTY	1992	ORGANIZATION SCIENCE	188
HAGEDORN	1993	STRATEGIC MANAGEMENT JOURNAL	180
TUSHMAN	1977	ADMINISTRATIVE SCIENCE QUARTERLY	180
BROWN	1995	ACADEMY OF MANAGEMENT REVIEW	179
HAGE	1973	ADMINISTRATIVE SCIENCE QUARTERLY	179

continued on following page

APPENDIX C CONTINUED

Lead Author	Yr.	Journal	Times Cited 2004
ORLIKOWSKI	1992	ORGANIZATION SCIENCE	179
COOPER	1987	JOURNAL OF PROD INNOVATION MGMT	173
MCFARLAN	1984	HARVARD BUSINESS REVIEW	172
ALLEN	1969	ADMINISTRATIVE SCIENCE QUARTERLY	169
ETTLIE	1984	MANAGEMENT SCIENCE	169
VON HIPPEL	1994	MANAGEMENT SCIENCE	165
HENDERSON	1994	STRATEGIC MANAGEMENT JOURNAL	162
GINZBERG	1981	MANAGEMENT SCIENCE	159
COOPER	1986	JOURNAL OF PROD INNOVATION MGMT	155
DEWAR	1986	MANAGEMENT SCIENCE	155
DESANCTIS	1992	ORGANIZATION SCIENCE	150
MAIDIQUE	1984	IEEE TRANS. ON ENGR. MANAGEMENT	145
LEONARD-BARTON	1988	RESEARCH POLICY	142
MARTIN	1983	RESEARCH POLICY	141
COOPER	1990	MANAGEMENT SCIENCE	139
CLARK	1989	MANAGEMENT SCIENCE	135
HOWELL	1990	ADMINISTRATIVE SCIENCE QUARTERLY	133
ANCONA	1992	ORGANIZATION SCIENCE	132
ABRAHAMSON	1991	ACADEMY OF MANAGEMENT REVIEW	130
MILLER	1982	STRATEGIC MANAGEMENT JOURNAL	124
PISANO	1990	ADMINISTRATIVE SCIENCE QUARTERLY	124
ROBEY	1979	ACADEMY OF MANAGEMENT JOURNAL	124
GUPTA	1986	JOURNAL OF MARKETING	119
ZIRGER	1990	MANAGEMENT SCIENCE	118
FREEMAN	1991	RESEARCH POLICY	117
MOWERY	1996	STRATEGIC MANAGEMENT JOURNAL	117
DAFT	1978	ACADEMY OF MANAGEMENT JOURNAL	115
THOMPSON	1965	ADMINISTRATIVE SCIENCE QUARTERLY	114
MONTOYA-WEISS	1994	JOURNAL OF PROD INNOVATION MGMT	113
MELONE	1990	MANAGEMENT SCIENCE	109
CLARK	1985	RESEARCH POLICY	107
MANSFIELD	1991	RESEARCH POLICY	103
ALDRICH	1972	ADMINISTRATIVE SCIENCE QUARTERLY	101

continued on following page

APPENDIX C CONTINUED

Lead Author	Yr.	Journal	Times Cited 2004
HOSKISSON	1988	STRATEGIC MANAGEMENT JOURNAL	99
LANE	1998	STRATEGIC MANAGEMENT JOURNAL	99
QUINN	1985	HARVARD BUSINESS REVIEW	99
ATTEWELL	1992	ORGANIZATION SCIENCE	98
MANSFIELD	1986	MANAGEMENT SCIENCE	97
BARLEY	1990	ADMINISTRATIVE SCIENCE QUARTERLY	96
MOED	1985	RESEARCH POLICY	96
DASGUPTA	1994	RESEARCH POLICY	94
PARSONS	1983	SLOAN MANAGEMENT REVIEW	94
SCOTT	1994	ACADEMY OF MANAGEMENT JOURNAL	94
VON HIPPEL	1987	RESEARCH POLICY	94
GREENWOOD	1996	ACADEMY OF MANAGEMENT REVIEW	92
MAIDIQUE, MA	1980	SLOAN MANAGEMENT REVIEW	91
MAHAJAN	1979	JOURNAL OF MARKETING	89
ULRICH	1995	RESEARCH POLICY	89
OLSON	1995	JOURNAL OF MARKETING	86
ROTHWELL	1992	R & D MANAGEMENT	86
SOUDER	1988	JOURNAL OF PROD INNOVATION MGMT	86
CHATTERJEE	1991	STRATEGIC MANAGEMENT JOURNAL	85
FRY	1982	ACADEMY OF MANAGEMENT JOURNAL	85
SANCHEZ	1996	STRATEGIC MANAGEMENT JOURNAL	85
GRIFFIN	1993	JOURNAL OF PROD INNOVATION MGMT	83
CHRISTENSEN	1996	STRATEGIC MANAGEMENT JOURNAL	82
NARIN	1987	RESEARCH POLICY	82
BAYSINGER	1989	ACADEMY OF MANAGEMENT JOURNAL	81
BETTIS	1995	STRATEGIC MANAGEMENT JOURNAL	81
DEWAR	1978	ADMINISTRATIVE SCIENCE QUARTERLY	81
ZMUD	1984	MANAGEMENT SCIENCE	81
HILL	1988	STRATEGIC MANAGEMENT JOURNAL	80
MILLSON	1992	JOURNAL OF PROD INNOVATION MGMT	80
LIBERATORE	1983	MANAGEMENT SCIENCE	61
WHEELWRIGHT	1992	HARVARD BUSINESS REVIEW	17

APPENDIX D

Table 2. Original articles sorted by increase in citations, 2004 to 2007

Lead Author	Yr.	Journal	Times Cited 2004	Times Cited 2007	Increase	% Increase
COHEN	1990	ADMIN SCI QUARTERLY	815	1685	870	106.7
MARCH	1991	MANAGEMENT SCIENCE	226	878	652	290.2
PRAHAL	1990	HARVARD BUSINESS REVIEW	590	1064	474	80.3
DAVIS	1989	MANAGEMENT SCIENCE	341	707	366	107.3
POWELL	1996	ADMIN SCI QUARTERLY	236	523	287	121.6
GRANT	1996	STRATEGIC MGMT JOURNAL	234	513	279	119.2
HENDERSON	1990	ADMIN SCI QUARTERLY	425	701	276	64.9
TEECE	1986	RESEARCH POLICY	487	752	265	54.4
TUSHMAN	1986	ADMIN SCI QUARTERLY	541	776	235	43.4
LEONARD-BARTON	1992	STRATEGIC MGMT JOURNAL	314	506	192	61.1
DAMANPOUR	1991	ACAD OF MGMT JOURNAL	249	438	189	75.9
DOSI	1982	RESEARCH POLICY	420	602	182	43.3
PAVITT	1984	RESEARCH POLICY	266	420	154	57.9
DESANCTIS	1992	ORGANIZATION SCIENCE	150	279	129	86
PORTER	1985	HARVARD BUSINESS REVIEW	306	432	126	41.2
DOUGHERTY	1992	ORGANIZATION SCIENCE	188	313	125	66.5
HENDERSON	1994	STRATEGIC MGMT JOURNAL	162	282	120	74.1
GREENWOOD	1996	ACAD OF MGMT REVIEW	92	211	118	128.3
BROWN	1995	ACAD OF MGMT REVIEW	179	295	116	64.8
VON HIPPEL	1994	MANAGEMENT SCIENCE	165	278	113	68.5
MOWERY	1996	STRATEGIC MGMT JOURNAL	117	222	105	89.7
ULRICH	1995	RESEARCH POLICY	89	191	102	114.6
BARLEY	1986	ADMIN SCI QUARTERLY	265	365	100	37.7
BANTEL	1989	STRATEGIC MGMT JOURNAL	222	317	95	44.6
DASGUPTA	1994	RESEARCH POLICY	94	185	91	96.8
SANCHEZ	1996	STRATEGIC MGMT JOURNAL	85	173	88	103.5
COOPER	1990	MANAGEMENT SCIENCE	139	227	88	63.3
MONTOYA-WEISS	1994	JOURNAL OF PROD INNOVATION MGMT	113	199	86	76.1
SCOTT	1994	ACAD OF MGMT JOURNAL	94	179	85	90.4
ABRAHAMSON	1991	ACAD OF MGMT REVIEW	130	214	84	64.6
ANDERSON	1990	ADMIN SCI QUARTERLY	190	274	84	44.2
CHRISTENSEN	1996	STRATEGIC MGMT JOURNAL	82	165	83	101.2

Lead Author	Yr.	Journal	Times Cited 2004	Times Cited 2007	Increase	% Increase
ANCONA	1992	ORGANIZATION SCIENCE	132	211	79	59.8
KIMBERLY	1981	ACAD OF MGMT JOURNAL	250	328	78	31.2
ABERNATHY	1985	RESEARCH POLICY	188	264	76	40.4
ORLIKOWSKI	1992	ORGANIZATION SCIENCE	179	252	73	40.8
PISANO	1990	ADMIN SCI QUARTERLY	124	196	72	58.1
MAHAJAN	1990	JOURNAL OF MARKETING	213	283	70	32.9
HAGEDORN	1993	STRATEGIC MGMT JOURNAL	180	248	68	37.8
HOWELL	1990	ADMIN SCI QUARTERLY	133	199	66	49.6
MANSFIELD	1986	MANAGEMENT SCIENCE	97	157	60	61.9
DEWAR	1986	MANAGEMENT SCIENCE	155	214	59	38.1
ATTEWELL	1992	ORGANIZATION SCIENCE	98	156	58	59.2
COOPER	1987	JOURNAL OF PROD INNOVATION MGMT	173	230	57	32.9
IVES	1984	MANAGEMENT SCIENCE	281	334	53	18.9
OLSON	1995	JOURNAL OF MARKETING	86	138	52	60.5
VON HIPPEL	1987	RESEARCH POLICY	94	146	52	55.3
NELSON	1977	RESEARCH POLICY	267	319	52	19.5
COOPER	1979	JOURNAL OF MARKETING	210	261	51	24.3
ZIRGER	1990	MANAGEMENT SCIENCE	118	168	50	42.4
MILLER	1982	STRATEGIC MGMT JOURNAL	124	174	50	40.3
BARLEY	1990	ADMIN SCI QUARTERLY	96	143	47	49
CLARK	1989	MANAGEMENT SCIENCE	135	181	46	34.1
CLARK	1985	RESEARCH POLICY	107	152	45	42.1
FREEMAN	1991	RESEARCH POLICY	117	161	44	37.6
GUPTA	1986	JOURNAL OF MARKETING	119	162	43	36.1
ETTLIE	1984	MANAGEMENT SCIENCE	169	212	43	25.4
TUSHMAN	1977	ADMIN SCI QUARTERLY	180	222	42	23.3
ROTHWELL	1992	R & D MANAGEMENT	86	127	41	47.7
LANE	1998	STRATEGIC MGMT JOURNAL	99	240	41	41.4
LEONARD-BARTON	1988	RESEARCH POLICY	142	183	41	28.9
MANSFIELD	1991	RESEARCH POLICY	103	142	39	37.9
ROTHWELL	1974	RESEARCH POLICY	204	243	39	19.1
GRIFFIN	1993	JOURNAL OF PROD INNOVATION MGMT	83	121	38	45.8
MELONE	1990	MANAGEMENT SCIENCE	109	145	36	33

Lead Author	Yr.	Journal	Times Cited 2004	Times Cited 2007	Increase	% Increase
ZMUD	1984	MANAGEMENT SCIENCE	81	114	33	40.7
CHATTERJEE	1991	STRATEGIC MGMT JOURNAL	85	118	33	38.8
COOPER	1986	JOURNAL OF PROD INNOVATION MGMT	155	188	33	21.3
GINZBERG	1981	MANAGEMENT SCIENCE	159	192	33	20.8
BETTIS	1995	STRATEGIC MGMT JOURNAL	81	113	32	39.5
MCFARLAN	1984	HARVARD BUSINESS REVIEW	172	204	32	18.6
QUINN	1985	HARVARD BUSINESS REVIEW	99	130	31	31.3
DAFT	1978	ACAD OF MGMT JOURNAL	115	146	31	27
DOWNS	1976	ADMIN SCI QUARTERLY	217	246	29	13.4
ZMUD	1979	MANAGEMENT SCIENCE	214	242	28	13.1
MAIDIQUE	1984	IEEE TRANSACTIONS ON ENGI-NEERING MANAGEMENT	145	172	27	18.6
HAGE	1973	ADMIN SCI QUARTERLY	179	203	24	13.4
SOUDER	1988	JOURNAL OF PROD INNOVATION MGMT	86	109	23	26.7
BAYSINGER	1989	ACAD OF MGMT JOURNAL	81	103	22	27.2
ALLEN	1969	ADMIN SCI QUARTERLY	169	191	22	13
MARTIN	1983	RESEARCH POLICY	141	162	21	14.9
FISHER	1971	TECHNOLOGICAL FORECASTING AND SOCIAL CHANGE	199	220	21	10.6
MILLSON	1992	JOURNAL OF PROD INNOVATION MGMT	80	99	19	23.8
HAGE	1969	ADMIN SCI QUARTERLY	198	217	19	9.6
MOED	1985	RESEARCH POLICY	96	114	18	18.8
HOSKISSON	1988	STRATEGIC MGMT JOURNAL	99	116	17	17.2
HILL	1988	STRATEGIC MGMT JOURNAL	80	95	15	18.8
NARIN, F	1987	RESEARCH POLICY	82	105	14	17.1
ROBEY	1979	ACAD OF MGMT JOURNAL	124	137	13	10.5
MAIDIQUE	1980	SLOAN MANAGEMENT REVIEW	91	102	11	12.1
FRY	1982	ACAD OF MGMT JOURNAL	85	91	6	7.1
THOMPSON	1965	ADMIN SCI QUARTERLY	114	126	6	5.3
WHEELWRIGHT	1992	HARVARD BUSINESS REVIEW	17	23	6	26.1
DEWAR	1978	ADMIN SCI QUARTERLY	81	86	5	6.2
MAHAJAN	1979	JOURNAL OF MARKETING	89	94	5	5.6
PARSONS	1983	SLOAN MANAGEMENT REVIEW	94	99	5	5.3
ALDRICH	1972	ADMIN SCI QUARTERLY	101	102	1	0.99
LIBERATORE	1983	MANAGEMENT SCIENCE	61	62	1	1.6

About the Authors

Robert S. Friedman is associate professor of humanities and information technology, director of the Science, Technology and Society program, and director of the Distributed Faculty of Social Sciences at New Jersey Institute of Technology, in Newark, New Jersey. His research examines science and culture, socio-technical systems design, and the role of technology in education. Dr. Friedman holds advanced degrees in the humanities and information systems, has published books and peer-reviewed articles concerning information and communication technologies, technologies in education, as well as the history of science and technology in 19th-century America. Dr. Friedman serves as editor-in-chief of the ACM's Special Interest Group for Information Technology Education's, peer-reviewed, *SIGITE Newsletter* and teaches an array of graduate and undergraduate courses on socio-technical systems in cultural contexts.

Desiree M. Roberts, PhD is an assistant professor and academic area coordinator of business, management and economics at the State University of New York-Empire State College. Her research interests are centered on innovation in services. She holds a Master's degree and a PhD in management from Rensselaer Polytechnic Institute. Dr. Roberts is a Rensselaer graduate fellow, and recipient of the New York State Chancellor's Award for Student Excellence and the New York State Senate's Women of Distinction Award. Additionally, Dr. Roberts is also President/CEO of Objective Analysis, Inc., a business management/operations consulting firm, specializing in turnaround planning, implementation and execution.

Jonathan D. Linton is the power corporation professor in the management of technological enterprises at the Telfer School of Management, University of Ottawa. Dr. Linton is the editor-in-chief of *Technovation: the International Journal of Technological Innovation, Entrepreneurship and Technology Management* and the director of the Emerging Technologies and Innovation Management Laboratory at the Telfer School. He also is on the editorial boards of *IEEE Transactions on Engineering Management* and *Technological Forecasting and Social Change*, and was formerly on the faculty of Rensselaer Polytechnic Institute and Brooklyn Polytechnic Institute. He has published over one hundred articles in academic and practice-oriented journals. Dr. Linton's research focuses on the challenges to operations associated to emerging technologies and innovation. Prior to entering academe, Dr. Linton consulted for a number of years and was an advanced manufacturing engineering at Ford Electronics Manufacturing. He has consulted to a variety of firms and governments in North America and Europe on issues related to the management and integration of technology. While Dr. Linton's research and work focuses on management of technology, he is still a registered Professional Engineer.

Index

G

H

I

J

K

L

M

N

O

P